ANIMAL LANGUAGE

ANIMAL LANGUAGE

Michael Bright

Cornell University Press

ITHACA AND LONDON

First published by Cornell University Press in 1985, by arrangement with the British Broadcasting Company.

First printing, paperback edition, 1985.

International Standard Book Number (cloth) 0–8014–1837–2
International Standard Book Number (paper) 0–8014–9340–4
Library of Congress Catalog Card Number 85-6714
Printed in the United States of America
Librarians: Library of Congress cataloging information appears on the last page of the book.

CONTENTS

INTRODUCTION

Two hundred years ago, an English clergyman, Gilbert White, conducted a correspondence with a naturalist friend debating the vexed question as to whether or not all owls hooted in the key of B flat. Doubtless, the issue, to his contemporaries, seemed a somewhat trivial one and I suspect that many people reading his Natural History of Selborne in the centuries since then have thought the episode no more than an amusingly eccentric episode in that captivating book. It was, however, one of the earliest instances of a scientific approach to animal communication. In the middle of the nineteenth century, Richard Burton, soon to become one of the greatest and bravest explorers of central Africa and a linguist of extraordinary accomplishment, studied a group of Indian monkeys and tried to compile a vocabulary of their language. He listed some sixty separate sounds; but he too abandoned the project.

That neither investigation proceeded very far can be attributed, at least in part, to the lack of the appropriate technological tools. Gilbert White had only a harpsichordist's pitch pipe and Richard Burton only the crude approximations of letters on a page. So in spite of the imaginative curiosity of such naturalists, the study of animal sounds did not progress very far. And so the situation remained until the beginning of this century.

Then, suddenly, a new tool was put in the hands of those who were interested in the subject. The gramophone was invented. One of the first to use it to record animal sounds was a German naturalist, Ludwig Koch. The practical difficulties with which he had to contend were daunting, for the early machinery needed to cut acoustic grooves in discs was extremely heavy and cumbersome—and alarmingly temperamental. Yet Koch took it to the most remote and uncomfortable locations and made pioneer recordings of mammals and insects as well as birds.

Koch was driven from Germany by Nazi persecution in the 1930s and came to this country. Here the BBC gave him financial support and regularly broadcast his recordings. His riveting enthusiasm for this still somewhat abstruse subject transcended the strange accents of his English and brought an awareness of the fascination and beauties of bird song to a large audience in this country. His recordings became one of the foundation stones of the BBC's library of natural history sounds.

Technological improvements since his time have been sweeping and swift. Tiny tape recorders, sonagraphs and many other devices and techniques have enabled investigators to record with comparative ease the most delicate of sounds, to make audible others that normally lie beyond the range of even the most acute of human ears, and to analyse in detail the most subtle and rapid variations of intonation and timbre. Scientists all over the world have taken advantage of the opportunities these tools have offered them; and the BBC's Natural History Library has continued to acquire and store the sounds they have captured.

In 1982, the Natural History Unit reached its twenty-fifth anniversary. It decided that one of its celebratory enterprises should be a survey of the latest discoveries in animal communication. The task of doing this fell to Michael Bright, one of the senior radio producers in the Unit. He travelled all over the world, talking to scientists both in the field and in their laboratories. On his return, he invited me to speak the narration that linked his recorded interviews. Both of us, I think, were astonished at the breadth and importance of what he had brought together. Happily, the substance of those programmes has not been lost, for he has recorded it in detail using that rather more ancient but still indispensible technique, the printed word.

Naturalists have, for centuries, laboured to provide us with superbly accurate detailed descriptions of what animals look like and what they do. But the discoveries described here so lucidly by Michael Bright, are of a quite different kind. This new breed of researchers approach the animals not from the outside but the inside. By deciphering what the animals themselves are saying to one another, they are giving us new and marvellously vivid insights into the very nature of the living creatures with which we share the world.

David Attenborough.

PREFACE

At two o'clock one cold winter's morning, our son Jonathan, just four weeks old, woke us from blissful sleep. Lying there for a few snatched moments, anticipating nappy change and feeding time, it occurred to me that his cries allowed this small and helpless creature to manipulate his parents to attend to his needs. His cry told us that he was hungry and uncomfortable. It could have indicated, by some subtle variation, that he was in pain or frightened. At the maternity hospital his mother recognised his cry as being quite distinct from the cries of other babies. In a simple cry there was information about his emotional state and his identity.

Curiously, it is not only the mother that responds to a baby's call. Even an eight-year-old child will experience an increase in heart beat rate, a rise in blood pressure and a change in skin conductivity, signs in the listener of anxiety provoked by the sound.

Crying is a baby's link with the outside world. When Jonathan cried he made things happen (and still does!), and in doing so learns about the world around him. By crying he elicits or not a parent's response, and begins to shape a concept of the environment, developing patterns of competence or helplessness that may persist throughout his life. He learns when calls are answered. Jonathan, like so many other animals, gets what he wants with sounds.

It was advances in recording and playback technology that really opened up the study of animal sounds: portable hi-fidelity tape recorders with sensitive, directional microphones, hydrophones which can record under the sea, computers that can synthesise sounds to order for experiments, detectors that extend our natural hearing range. Two decades ago, researchers would have annotated sounds as phonetic shapes or musical notations. Today, the sound spectograph turns sound into pictures which we, as predominantly visual animals, find much easier to interpret.

Fifty years ago, who would have foreseen the discovery that bats, navigating by sound signals, can fly at 40 mph in the dark and pass cleanly between black threads set only eight inches apart? How could we have guessed that a dolphin is able to pinpoint an object, and perceive its nature and structure, simply by beaming sound at it? Who would have thought that the nasal splutterings of bull elephant seals or the drumming of woodpeckers would show local dialects?

Without the technology, scientists would have little to say. With it, sounds can be recorded and analysed and we may begin to interpret their meaning. After the initial flood of discovery comes the trickle of understanding. Why are some birds' songs so elaborate while others are not? Do animals learn their songs and calls or are they innate? What triggers a bird to sing, a wolf to howl, or a cricket to chirp? How does an animal filter out background noise and chatter and hear only relevant signals? Does an animal expose itself to danger when it calls? Why do birds sing at dawn? Can we learn about our own language by studying the sounds animals make? We can provide answers, at least in part, to many of these questions, although the question of whether animals have language remains unresolved.

This book deals with only one medium of communication: sound. I have ignored vision, touch and smell. But even if we only consider sound, we can see that animals have developed remarkable communicative abilities – the speed with which a cicada vibrates its tymbal, the rapid transfer of information in dolphin clicks, and the distance over which a humpback whale song or an indri's 'loud call' may travel. These communication systems may or may not constitute a language but they are so fascinating in themselves that the question seems largely academic. The facts themselves are awesome enough.

While preparing this book, and the series of radio programmes on which it is based, I and my colleagues visited 50 research establishments and universities throughout Europe and North America and talked with over 100 research scientists. I hope we have represented accurately their achievements and aspirations. To a large extent, their interests have determined the contents of this book – as is inevitable in any report from the front line of an expanding field of study. The results are fascinating both in detail and in their implications; but, at this stage, the research is still concentrated into relatively few groups of animals, and into relatively few species within those groups. The emphasis is reflected in this book. Those who wish that it were not can take consolation from the reflection that, within the next few years, other animal communication systems will have been studied which will be every bit as remarkable as those described in the following pages.

1
COMMUNICATION

The natural world is never silent. Plants move in the wind; the surf crashes onto the sea shore; the wind blows through caves and crevices. Animals, including man, make a major contribution to the cacophony, sometimes just by brushing through the undergrowth or splashing in the sea, but more often by their deliberate calls and songs. Man, inevitably, has invaded even this natural domain. Noise pollution is as serious to a natural ecosystem as an oil spill or acid rain. How much, for example, has the deep engine throb of bulk carriers or supertankers upset the communication systems of the great whales, and contributed to their near extinction? We shall never know.

What we *do* know is that the sounds made by insects, crustaceans, spiders, fish, amphibians, reptiles, birds and mammals are the product of natural selection and help the animals to survive. The human baby or wolf cub, the superb lyre bird or the lion, can use sound to attract, inform, or warn. They can inform others where they are, what state they are in, where food may be found, that they should not move nearer, that they may approach closer, that they are stronger or weaker, or that they are closely related. Sound can also inform predators so it must be carefully controlled. It may be in competition with the sounds of others. To avoid interference and confusion in a noisy environment sounds need to be highly specific and used in particular behavioural contexts.

By far the most common use for sound, then, is to communicate, but what exactly is communication? For one thing, it is notoriously difficult to define to everyone's satisfaction. However, I will settle for a simple definition: an animal has communicated with another when it has transmitted information that influences a listener's behaviour.

If a blackbird should see a cat in the garden it gives an alarm call, and all the other birds in the neighbourhood flee for cover. The bird that has spotted the cat has conveyed information to the rest, which were able to take evasive action. Each bird's behaviour was influenced by the sound. But who is benefiting from such behaviour?

The birds in the area benefited. They received an early warning that a cat was about. The spotter benefited too – it escaped in the mass confusion. On the surface, it seems that all have benefited (all except the cat, that is), but is this the case? Modern approaches to animal behaviour warn researchers to beware of mutual benefits. If a bird spots a cat and gives a cry, it draws attention to itself and may increase the risk of being caught itself. A bird that cheats by remaining silent would not run this risk, but it might take advantage of other birds that do cry out. There must be direct benefit to the bird that sounds the alarm. There is also benefit to the recipient or it would not respond to the call. It is argued sometimes that a particular piece of behaviour evolves because it benefits the genes of the behaving animal, but it takes two to communicate. Are, then, both beneficiaries, or just the sender, or just the receiver? As in many areas of animal behaviour, debate is heated.

Professor Peter Marler, at the Millbrook Field Station of Rockefeller University, New York, feels that both sender and receiver benefit, to some extent, in an act of communication. He and a number of other researchers have tried to suggest that somewhere there is the key notion of mutuality, although sometimes the rules may be broken and an individual may lie and gain benefit as a consequence. Lying, though, suggests Marler, is a peculiar phenomenon in that it implies a defiance of rules. There have to be rules to be broken. At present he feels that researchers are unlikely to make significant progress in understanding lying and subsequent manipulation until a more complete picture is gained of the truthful mode.

Dr Peter Slater, of the University of Sussex, thinks that evolution will favour the sender. He feels that although in many cases it may be correct that both sender and receiver benefit, the general rule that dictates whether or not a communication evolves is that there should be benefit to the sender; the receiver doesn't matter as far as evolution is concerned. Slater cites the example of distraction displays in wading birds as not transmitting useful information to the receiver – the wader is lying. If the receiver understood that it was being deceived it would ignore the signal.

Dr John Krebs and Dr Richard Dawkins, of the University of Oxford, go even further. They propose that communication evolves because it enables the sender to influence profoundly the behaviour of the receiver. Signals from actors manipulate the behaviour of reactors. At the same time there is the evolution of resistance on the part of the receiver who is being manipulated. Krebs draws on the cuckoo nestling being fed by foster parents as an example of this continuous evolutionary tug-of-war. He feels that it is not helpful to consider the young cuckoo's gaping mouth as communicating something useful to the sedge warbler. Rather, the cuckoo is manipulating the sedge warbler's behaviour, to the advantage of the cuckoo and to the disadvantage of the warbler parents. During the course of evolution there has been an evolutionary race between sender and receiver. At this point in time, the cuckoo is leading. In, say, 10,000 years the sedge warbler may have developed the capacity to discriminate against cuckoos.

Nevertheless, in many cases both parties do benefit, and although they may not share the benefits equally, it may be that the balance of selective forces acting on each is, in fact, pushing them in the same direction. Courtship signals fall into this category; so, too, do signals exchanged between members of an ant or bee colony, or a wolf pack; and there are the signals between mother and baby.

In a seabird colony, parents and chicks call to each other. The chick signals its identity, and is recognised by its own parents from the thousands of other young birds in the colony. The same sound signals may have another meaning, such as 'I want food'. If disturbed by a predator the chick will change its signal to one of alarm. Then there are signals for threat, attack, retreat, aggression, submission, or even plain friendship. The 'mewing' call of the lesser black-backed gull *Larus fuscus*, for example, is a long drawn-out note produced with the bill stretched forward – it indicates friendliness towards a mate or young in the nest in a kind of greeting ceremony. On the other hand, there are calls that are not so friendly. A threat call advertising ownership of a nest site may eventually lead to a fight. Seabirds also have courtship and mating signals that identify partners and help cement the bond between them.

Sound, though, is not the only communication system available. There is the chemical communication of taste and smell, like the pheromone system used by moths; and there is the tactile communication of two monkeys taking part in mutual grooming. There are visual signals, like the leg waving of spiders, the flashing of fireflies, or the spectacle of the peacock's tail. Why then is sound so important? What are its advantages?

Peter Slater believes that a primary reason for using sound is that an animal can get far more information packed into a signal for a given unit of time. If a conversation depended on, say, smell, it would be difficult to get rid of the first part of a message in order to send the next bit. There is little a sender could do to manipulate the signal, and it would be extremely laborious to send a long message. The same is true of visual signals. The rapidity with which even a semaphore-type of signal can be changed does not compare with the speed with which a sound signal can be turned on and off. Sound can also go around corners. At night sound and smell have obvious advantages over sight (unless an individual can make its own light) but scent is likely to linger and attract predators. Sound also enables a creature to announce its territorial rights without having to visit all parts of its domain.

In the main, though, animals have developed a combination of communication channels, using the most appropriate for the task involved. Ring-tailed lemurs *Lemur catta*, for example, have long black-and-white tails for visual signalling, a large repertoire of sounds, and special glands for scent marking and 'stink-fights'. Is sound, then, more important than visual or chemical cues? There is much argument amongst those studying child development, for instance, about whether sound, vision, touch or smell is of primary importance. In some cases, however, sound is clearly at the top of the hierarchy.

In experiments with turkeys, a deaf female laid eggs, sat and incubated them normally, but behaved in a very odd way when they hatched. She was unable to recognise the chicks and proceeded to kill them. Normally the youngsters' calls seem to induce recognition and suppress the mother's natural aggression. Despite visual information the deaf mother was dominated by her aggressive behaviour.

Similarly, in tests with chickens, a chick was isolated from its mother. It was placed in a bell jar and could not be heard. Although the hen could see her chick she ignored it and walked away.

In a series of playback experiments with bush crickets, a female preferred to head in the direction of an inanimate loudspeaker, although she passed by several desirable, but silent, males standing right next to her.

So what is this thing called sound, which can play such an important role in animal behaviour? The simplest sounds are continuous pure tones like the sound of a tuning fork or the sine wave signal from a synthesiser keyboard. Biological sounds tend to be more complex with overtones or harmonics (multiples of the frequency of the fundamental note) spread across the sound spectrum. Each creature produces sounds with a distinctive quality or timbre. In the same way that a particular note played on a piano sounds different from the same note played on a trumpet, so, too, do the notes emitted by a gibbon and a cockatoo. A gibbon calling middle C will produce the fundamental plus a set of harmonics. The cockatoo will produce the same fundamental but a different set of harmonics.

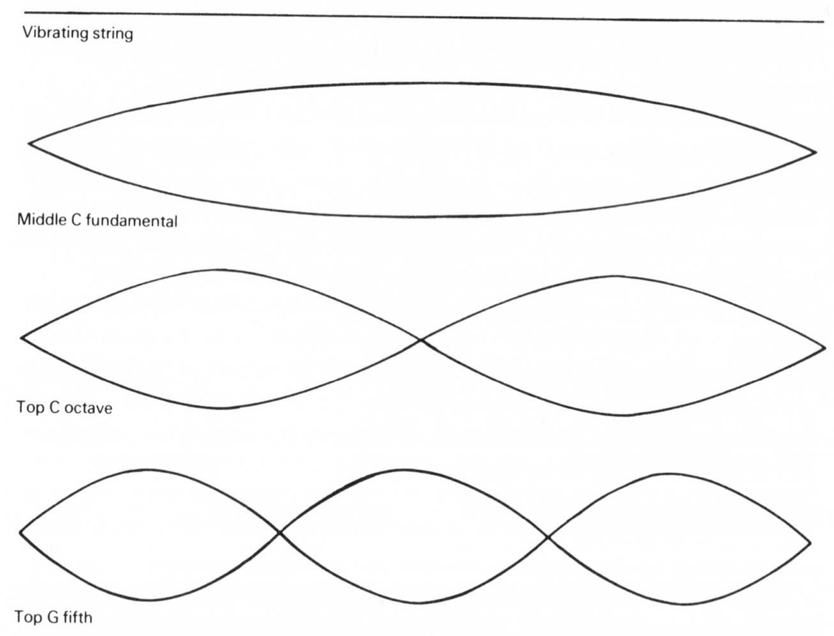

Fig. 1 Vibrations in a violin string.

Sounds are waves of alternating pressure changes that pass through a medium. The medium can be a gas, like air, a liquid, like water, or a solid, like a tree trunk. The only thing that cannot conduct sound is a vacuum. Sound waves are formed when pressure changes spread from the sound source, much like ripples spread on the water surface when a stone has been thrown in. They can be produced by striking, rubbing, scraping, or by blowing and sucking.

The intensity or volume of sound is measured in decibels (dB), a logarithmic unit for comparing two amounts of power. It reflects the fact that sound waves carry energy. A whisper, for example, has an intensity of about 30 dB, human conversation is 60 dB, and a jet aircraft heard about 30 metres away is at 140 dB. The human pain threshold is at about 120 dB.

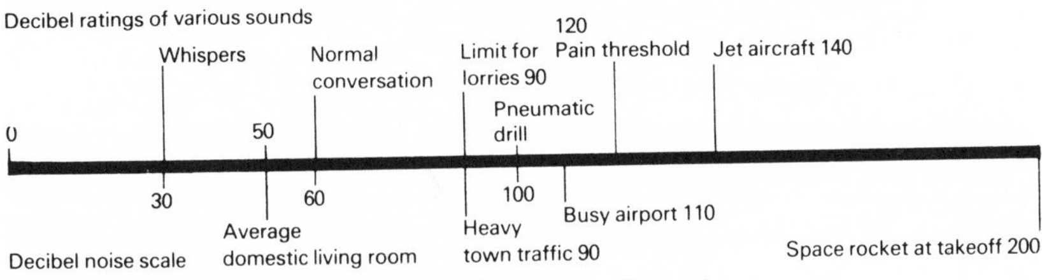

Fig. 2 Sound levels in decibels (dB) of a variety of sounds.

The wavelength is the distance between one pressure peak and the next, and the frequency is the number of waves that pass in one second. One complete wave cycle passing per second is known as one Hertz (Hz). The longer the wavelength the lower the frequency, and the lower the note that is heard.

The bottom note on the piano has a frequency of 30 Hz or 30 cycles per second. The highest note has a frequency of 4.1 kilohertz (kHz) or 4,100 cycles per second. Middle C is 256 Hz. The average human can hear sounds between 20 Hz and 20,000 Hz (20 kHz) and we are especially sensitive to sounds at around 2,000 Hz. (Throughout this book I have kept frequencies in Hertz for easy comparative reference.)

Some animals can make use of sounds at very high frequencies, far beyond the range of human hearing. These are ultrasounds. Bats, for example, can produce and hear sounds right up to 200,000 Hz. At the other end of the scale are the very low frequency rumbling noises, known as infrasounds. Pigeons have been reported to be able to detect infrasonic noise down to less than one Hertz.

In order to have an easily identifiable signal most animals vary or modulate the frequency; in other words, they change the pitch. The song of the blackbird is a good example of a frequency modulated signal.

Apart from frequency, or pitch, animals can also vary the volume and the temporal patterning or spacing-out of their sounds. Volume is related to size – large animals make loud noises, but some work hard to overcome this

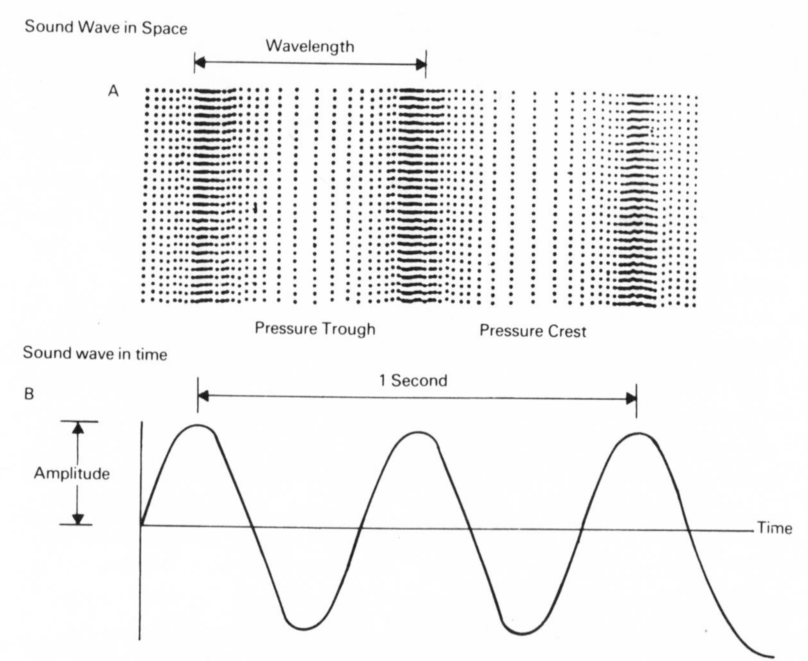

Fig. 3 Two views of a sound wave: A. Alternating phases of compression and rarefaction. B. Sound wave passing a stationary observer. Two complete cycles in one second gives a frequency of two hertz (2 Hz).

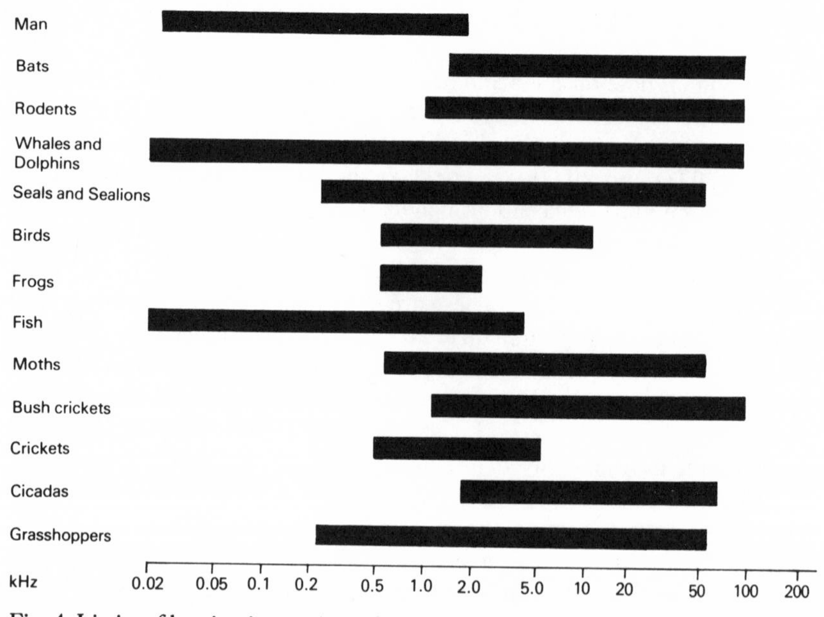

Fig. 4 Limits of hearing in a variety of animals.

limitation. One species of mole cricket *Gryllotalpa vineae* digs a burrow that is shaped like a twin-horned megaphone. Sitting at the mouth of the horn the creature can blast out a song that is an uncomfortable 90 dB a metre away. On a quiet night this five centimetre-long insect can be heard 600 metres away across the countryside.

The cricket also tells us something else about a sound signal – pattern is important. Crickets (Gryllidae) sometimes vary the pitch of individual notes, but that appears not to matter. The female is listening to the pattern of notes, the temporal organisation. It is the spacing, and not the frequency, that distinguishes one species from another.

Complex sounds that make use of changes in pitch, loudness, and pattern can convey far more information than simple tones. They reach their peak of sophistication in human speech, whale songs and the beautiful songs of birds. How, though, do each of these creatures produce their sounds?

Vocal animals blow air through an orifice or across membranes, and instrumentalists rub appendages together or beat whatever they are standing on. Mammals and birds use air from the lungs to make special membranes vibrate. In mammals, for example, sounds are generated as air is passed over the vocal chords in the 'voice-box' or larynx. In humans, the quality of sound is further modified as it resonates in the cavities within the head, and is manipulated by movements of the tongue and mouth. Just as people have differently shaped bodies and faces, so too each human being has a different voice. In birds, the syrinx, the equivalent of the 'voice-box', is at the junction of the bronchi and is in two halves so that some birds can sing two songs at the same time. Some frogs and toads, howler monkeys and siamangs, have resonating air chambers that help amplify the sound signal.

Insects, spiders and crustaceans are instrumentalists. They produce sounds by tapping, scraping, bending, clicking or rasping parts of their exoskeleton. Grasshoppers (Acrididae) scrape a comb on the hind leg across the surface of the front wing. The lesser water boatman *Corixa punctata* rasps a series of bristles inside the foreleg against the head. The sound is amplified by air sacs on either side of the head. Bugs, like cicadas (Cicadidae) have a skeletal plate on either side of the body which is buckled in and out like a tin lid. The screech beetle *Hygrobia hermanni* rubs the end of its abdomen against a file beneath the tips of the wing cases or elytra. The death's head hawk moth *Acherontia atropos*, on the other hand, joins the vocalists in that it produces a clearly audible squeak by forcing air through the proboscis, very much like a clarinet. Click beetles (Elateridae) unexpectedly snap to startle a predator, and bombardier beetles *Brachinus crepitans* explode, discharging acrid scent, accompanied by a loud report. The death's head caterpillar makes a series of loud snapping noises much like electrical sparks, when handled. The death-watch beetle *Xestobium rufovillosum* makes its insistent vibration by inclining forward and bashing the underside of its body against the wood.

Fish are noisy. They make sounds by scraping or vibrating the swim

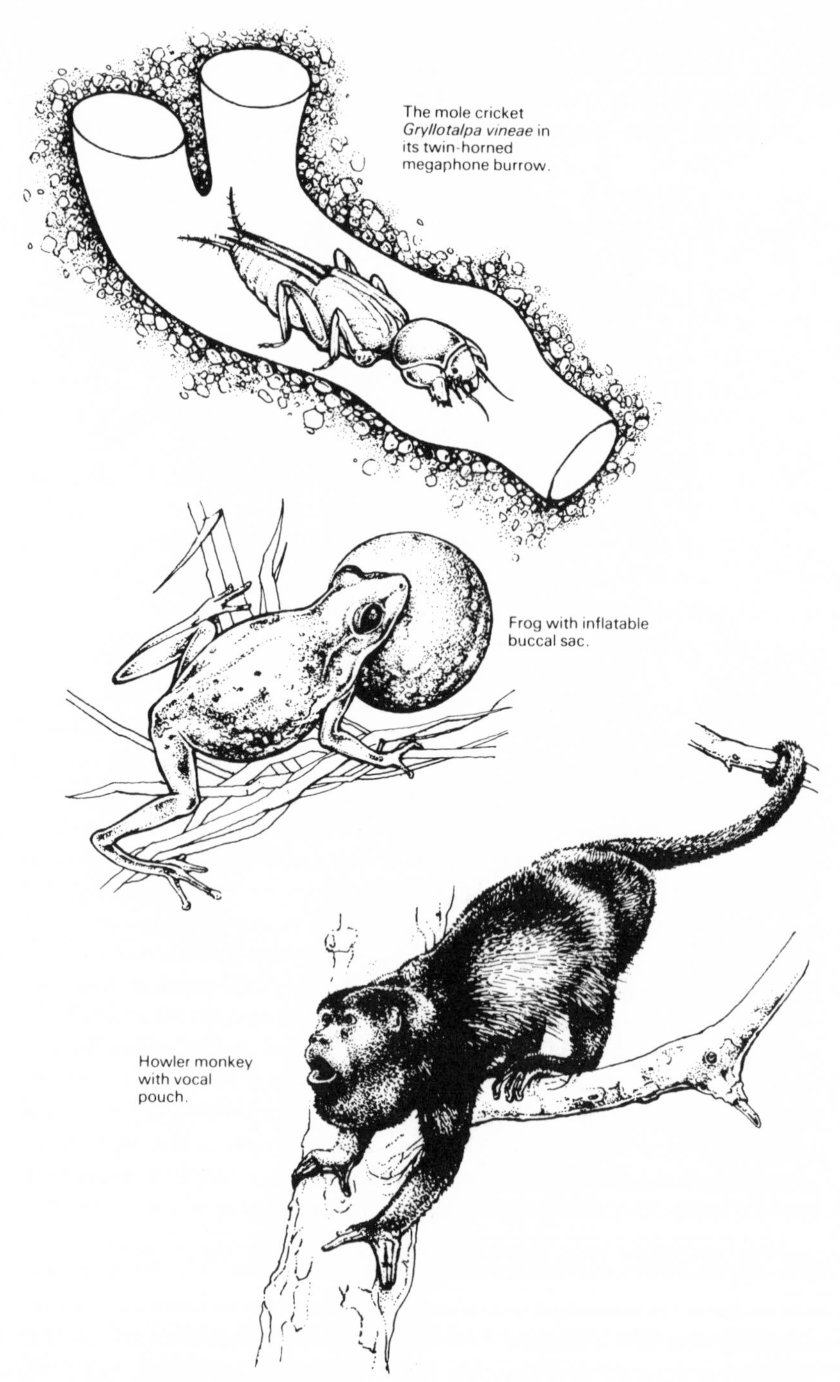

The mole cricket
Gryllotalpa vineae in
its twin-horned
megaphone burrow.
Frog with inflatable
buccal sac.
Howler monkey
with vocal
pouch.

Fig. 5 Sound resonators in animals.

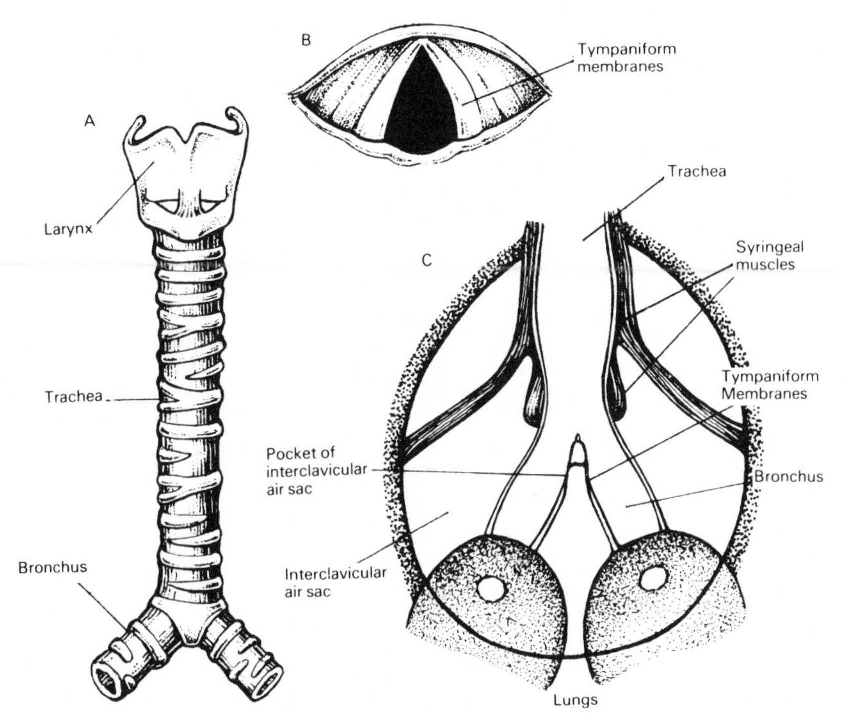

Fig. 6 A. The human larynx. B. Cross section of the larynx. C. Vertical section of the bird syrinx.

bladder, or by grinding their teeth, or by plucking special spines. Rattlesnakes *Crotalus spp.* despite being deaf to airborne sounds themselves, have specially modified scales in the tail with which to rattle out a warning. A desert gecko *Teratoscincus scincus* has a stridulatory organ on the upper side of the tail. A series of nail-like scales rub against each other to produce a cricket-like sound.

Some birds, such as storks (Ciconiidae), clap their beaks together, and snipe *Capella spp.* make a fluttering or 'bleating' sound as they descend out of the sky. As they fall rapidly through the air with the tail expanded, two modified feathers sticking out either side of the tail vibrate in the air flow. The American woodcock *Scolopax minor* precedes its flight call, during roding, with a trilling sound produced by modified outer primary feathers in each wing. The broad-tailed humming bird *Selasphorus platycerus* similarly produces a trill with its high-speed wings. Woodpeckers (Picidae) hammer out a rapid pattern of sounds on hard resonant surfaces, such as tree-trunks.

Gorillas use their chests as drums, while many other primates shake bushes and beat sticks on the ground to underline a display with sound. Rabbits *Oryctolagus cuniculus* stamp their large hind feet and beavers *Castar spp.* have been heard to smack the water with the broad, flat tail. Porcupines (Hystricidae) and tenrecs (Tenrecidae) rattle their spines.

In order for communication to take place, the receiver must be able to hear the sound. There must be an organ for converting sound waves into vibrations, and in turn into nerve impulses which represent the original signals and register in the brain. Such an organ is the ear, but ears need not be on the head. Crickets hear through their elbows, spiders detect vibrations through their feet and fruit flies hear via their antennae.

The ears of small insects and fish tend to be of a type known as particle displacement ears. A sensitive hair, small enough to be moved by the displacement of air or water is connected to a nerve that translates the movement of the hair into impulses along the nerve. Larger animals have ears that are pressure receptors. A membrane vibrates in sympathy with the sound and nerves again translate the vibration into electric impulses.

To make the most of low-level sounds, many animals have a funnel (the external pinna in humans, often loosely called the ear) to collect the sound from a larger volume of air and channel it down to the eardrum. Prey animals and nocturnal creatures, especially, have sensitive mobile ears that can be swivelled around to scoop up sound from any direction. The desert fox *Fennecus zerda*, for example, has large ears that allow it to hear food scrabbling about below the surface of the sand, and can detect an approaching predator many miles away. It has been found that certain crickets and owls have the equivalents of external pinnae.

Sound is ubiquitous in animal life, but the very simplest forms appear not to use it. How did the use of sound evolve? Signals probably evolved from non-signal movements which originally served different purposes. The change to a communication function is a type of process called ritualisation.

Anything that helps an animal to send or receive useful information will be favoured by natural selection. The receiver will become more sensitive to the informative sounds while the sender may, if it is to his advantage, become better at sending the signals. A barely detectable sound, say the gasp of an intake of air, might signal to a listener that an animal is getting ready to attack. The listener will become more sensitive to the sound. More important, the attacker might deliberately make the noise in order to signal attack, and here we have the beginnings of a threat signal. Emotionally induced changes in rate of breathing may have been the starting point for the evolution of specialised structures which would enhance the sound. The signal would be made louder and more controllable with the development of, say, special membranes in the respiratory passage, culminating in the highly sophisticated larynx or syrinx.

The insects followed another route. An insect inevitably makes a noise when it moves. The wings or legs of crickets and grasshoppers must have rubbed together. Ears may already have developed for predator detection. Over millions of years it would be an advantage to females to recognise the sounds of males and an advantage to the males to amplify them.

What were the first sounds? Which was the first animal to break the silence? Was it, perhaps, a noisy oyster clapping its shells together meaningfully

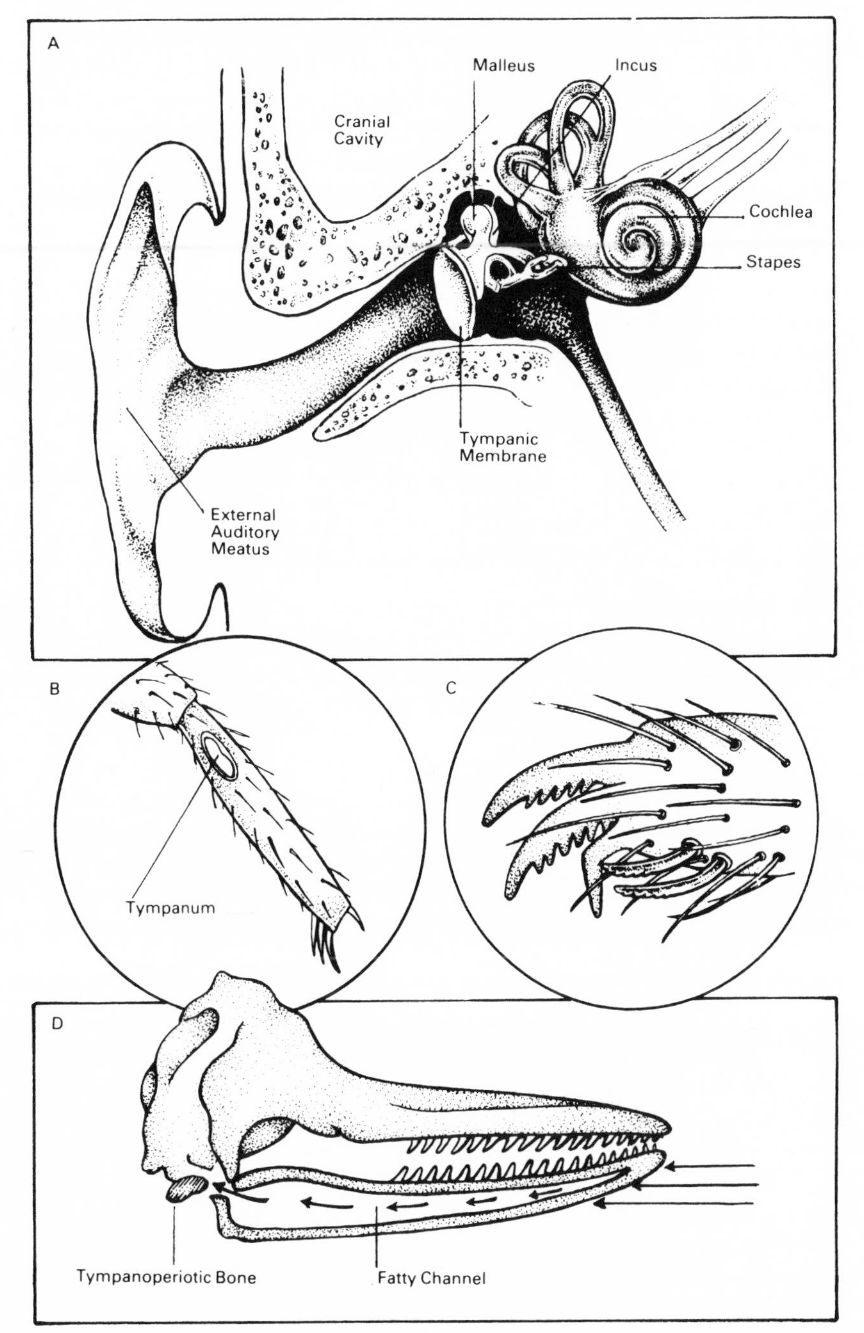

Fig. 7 Animal ears: A. Generalised mammalian ear. B. Insect ear. C. Spider's foot. D. Dolphin receives sound via lower jaw.

beneath the waters of the Triassic seas nearly 250 million years ago? Or was it on land; perhaps the rustling of an insect's leg, the croaking of an amphibian, or the hiss of a dinosaur?

All these are possible, but I believe there is an even better candidate. Trilobites, which lived on the bottom of the oceans 500 million years ago, were a very successful group. They had hard exoskeletons admirable for making sounds, and were around long enough to have evolved complex sound-making equipment. Many of them, according to the fossil record, had excellent eyes, which may suggest that they did not need sounds to find or recognise things. But what if, like their distant extant relatives the horseshoe crabs, they mated at night? Could it be that they produced specific signal noises to help find each other in the darkness? We shall probably never know, although someone, some day, may find a trilobite with structures that can only be interpreted as organs for producing or receiving sound.

2

SINGING WHALES

Humpback Whales

At the turn of the century, humpback whales *Megaptera novaeangliae* became one of the main targets of the voracious international whaling industry. These playful whales were easy pickings for whaling ships. Humpbacks grow to 50 ft long and weigh 40 tons or more. Fat content is lower than in other large whales but they are easier to find and catch. Their regular and easily located congregations for breeding in winter, and feeding in summer, meant that they were more exposed to hunting than the solitary blues and fins. In Antarctic waters alone an estimated population of 22,000 southern humpbacks was reduced to barely 3,000 individuals by the early 1960s when, thankfully, a ban on humpback whaling was introduced. The humpback whale, though, has a global distribution with migrations to the food rich Arctic or Antarctic waters in summer, and to the breeding grounds around the tropics in the winter. In summer the northern race of humpbacks spends its time in the Arctic, filtering out small fish and shrimp-like crustaceans called krill. Apart from a few social interactions when meeting others, they are usually silent. It is after their migration south, to breed and calve in warmer waters, that their beautiful sounds are heard. In the Atlantic ocean, groups of whales can be spotted off the islands of the West Indies, around Bermuda and off the west coast of Africa. In the Pacific, southern California and Hawaii are known gathering places. This seasonal predictability and easy discovery nearly led to their extinction. Since the turn of the century the worldwide population is estimated to have been reduced by 95% by whaling. Humpbacks are currently protected throughout their ranges by International Whaling Commission regulations. Except for limited aboriginal kills in the Caribbean and Greenland and inadvertent deaths when being caught in fishing nets off the east coast of North America, they are now safe. Ironically it was the whalers who first noticed that their prey was sensitive to sound. The disturbance of a noisy oar in its rowlock would frighten a whale away. There is even record of an underwater camera shutter release scaring a humpback whale 20 metres away.

Their protection came not a moment too soon, for extinction would have robbed us of a remarkable animal now known worldwide for its beautiful underwater songs. The songs were first recorded by O. W. Schreiber in 1952 from a US Navy underwater listening post on the submarine slope of Kauai, Hawaii, but he didn't recognise which marine creature was responsible. William Scheville at Woods Hole Oceanographic Institution later identified the sounds as coming from humpback whales. The humpback subsequently became a world bestseller on an LP record called *Songs of the Humpback Whale*. The scientist responsible for drawing scientific attention to their songs is Dr Roger Payne of the New York Zoological Society who, together with his wife Katy, a musicologist turned zoologist, has been involved in an intensive study of humpbacks, their songs and behaviour. Roger Payne's interest in bio-acoustics began with the directional sensitivity of bats' ears; continued with the owl's ability to locate prey by hearing; and then followed with studies of the way some moths can avoid bat sonar. His first encounter with whales was on a rain-washed beach near Tufts University where a small porpoise had been stranded. Payne was appalled by what he saw.

> 'It had been mutilated. Someone had hacked off its flukes for a souvenir. Two other people had carved their initials deeply into its side, and someone else had stuck a cigar butt in its blowhole. I removed the cigar and stood there for a long time with feelings I cannot describe.'

He resolved at that moment to learn about whales, so that some day he might be able to have at least some effect on their future.

The opportunity came in 1967 when Roger and Katy Payne visited Bermuda to look at migrating whales and met Frank Watlington, an acoustics engineer at Columbia University Geophysical Field Station. Watlington was responsible for recording anything whether natural or man-made that produced sounds in the sea, in particular under-sea explosions. He played them underwater recordings of humpbacks and lent his tapes. While out in a row-boat, attempting to get close to humpback whales without frightening them, Roger Payne heard the sounds of whales amplified through the bottom of the boat. He noticed that one of the whales was singing phrases similar to those on Watlington's tape, and it suddenly became apparent to Payne that he was listening not to random sounds but to regular repeating patterns.

The humpbacks make a variety of grunts and squeaks too – indeed, they appear to have a whole repertoire of vocalisations – but these repeating patterns are by far the most interesting. They could be considered as true songs for they consist of long complicated repetitive sequences, much like bird song except that unlike most bird songs, each one can last from five to more than 30 minutes. (It even resembles bird song if speeded up.) A whale might stop singing or resume singing at any point in the song so it is difficult to determine a beginning, middle or end. Each song may be sung over and over again without breaks for many hours. A whale recorded in the Caribbean by Howard and Lois Winn, of the University of Rhode Island, sang non-stop for 22 hours, breathing in the intervals between phrases. It was still going strong when the Winns pulled up hydrophones and went home. Humpback whale songs are the longest and most complicated of animal songs known to man. Day and night the Paynes collected their sequences of whale songs by dangling hydrophones or underwater microphones from outriggers to each side of their small sailboat. The sounds were then recorded on tape-recorders and taken back to the laboratory and analysed with the help of Scott McVay at Princeton University. Together with Frank Watlington's recordings they have an almost continuous record of the Bermuda whales for a period of over 20 years.

In an attempt to analyse the sounds, the Paynes divided each song into identifiable parts. The smallest part is a unit (equivalent to a musical note). Units are grouped into small repeating sequences called phrases. Groups of similar phrases are known as themes. Off Bermuda, eight to ten themes make a song (later off Hawaii, as few as four to five themes were found to make a song), and the song is repeated without pause as a song pattern. By breaking down the whale sounds in this way the Paynes were able to make a series of interesting observations.

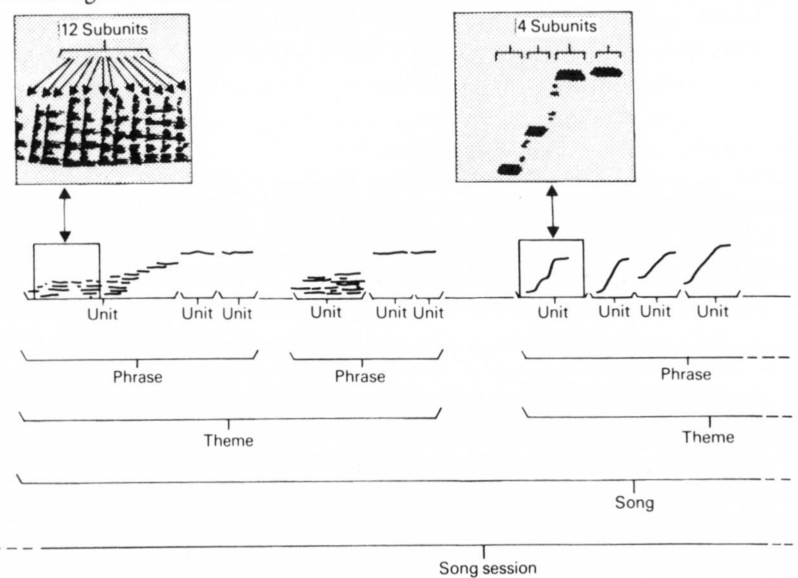

Fig. 8 Simplified humpback whale sonagram showing subdivisions of songs. (Insets show sections expanded if slowed down).

It turned out that all humpback whales in a particular area sing the same song. Whales from the Pacific Ocean, however, sing a different song from those in the Atlantic, although the laws on which they base their songs appear to be the same in both populations. A song, for example, might consist of five themes (A-B-C-D-E) which always follow in the same order. If a theme, say C, is dropped, the remaining themes are always sung in the same sequence (A-B-D-E). In much the same way, the laws that govern the structure of sonnets, or the composition of western music are the same whether the piece is composed in New York or London. The laws governing humpback whale songs are, however, complicated and little understood and it is not known whether they are inherited culturally or genetically.

The migration routes of the Atlantic and Pacific populations are unlikely to cross, hence the dialect-like differences. But there are separate centres of whale breeding activity in the Pacific Ocean. Baja California and Hawaii, 3,000 miles away, for example, are two separate breeding areas. The question is, do whales from each area meet up and share similar songs or are there dialect differences even between Pacific whales? A photographic technique identifying distinctive markings on the underside of humpback flukes, used by James Darling of the University of California at Santa Cruz, together with song analysis, helped to give an answer.

A study in 1979 revealed that Baja and Hawaii humpback whales were singing the same song. Two years later a follow-up showed them to have made identical modifications to the song. The whales from separate breeding areas must have been in contact, for songs could not physically travel 3,000 miles from one breeding ground to the other. Analysis of fluke-pattern photographs confirmed that whales feeding off south-east Alaska would appear in the breeding areas in both Hawaii and Baja California. The degree of interchange between these areas is still unknown.

This discovery had other implications. It meant that whales might visit any one of several breeding areas, a factor important for conservation. If, one year, a breeding area was affected by some man-made or natural catastrophe and all the whales killed, whales which that year had visited an alternative area might turn up after the site had recovered and repopulate the area. Also, when counting whales to establish stocks and whaling quotas, it would be quite possible to count the same whale twice – once in each breeding area – and get an over-estimate of numbers. By studying whale sounds the researchers had identified a conservation loophole.

Another surprising discovery was that the songs are continually changing. All the whales in an ocean population change their songs in the same way so that each individual is up to date with the current vogue – a kind of top-of-the-pops. The changes are progressive and rapid, each component perhaps changing every two months, although many elements of the song are changing at any one time. Taking a folk-song analogy, in the first month an individual might sing: 'London's burning, London's burning. Fire-fire, fire-fire . . .' and so on, while two months later one phrase would be modified so the song becomes: 'London's burning, London's burning, London's burning, Fire-fire, fire-fire . . .' etc.

Humpback whales are composers, much as humans are, except that the whales do not create new songs, they evolve something different from what they already have by making minor modifications. During a season a song may change components, add extra parts and drop in pitch. A component may undergo rapid changes for several months and then be left alone while other parts of the song are modified. Curiously, newly created phrases are sung more rapidly than older ones. Sometimes a new phrase is created by taking the first and last parts of older phrases and dropping the bit in the middle, much as we shorten 'I would' to 'I'd'.

What selective advantage a whale may gain from changing its song is unclear. Whether it is a dominant trend setter who introduces the changes will be difficult to find out, but what is clear is that an entire song is renewed totally after about eight years.

Humpback whales only sing on the breeding grounds and occasionally during the migration. Only one isolated and misguided individual has been heard singing in the feeding areas in the Arctic or the Antarctic. What happens to the song, therefore, between breeding seasons? Whales, like elephants it seems, never forget. In the new season the song is picked up usually as it was left the previous season, remembered by each male in the population. Most of the changes that take place occur during the singing season and not in between. (There is only one case of a substantial change in a theme, apparently between seasons.)

With such sophistication in song development, the question often asked is whether whales are highly intelligent creatures capable of indulging in intellectual pursuits and communicating in complex languages, even to the point, as some have suggested, of being wiser than man. True, whales have large brains, in fact the largest brains of any animals that have ever lived. How they use this large brain is unknown. It is tempting to suggest that whales reason about the world just as man does, but they are constrained by the limits of the environment in which they live. Whales lack hands and therefore brainpower cannot be channelled into the creation and use of tools, for example. They do, however, need to develop senses and skills to maintain their efficiency under the sea. Perhaps a part of their large brains is used in running this sophisticated sound system capable of discriminating songs above the backyard jungle of the noisy ocean, interpreting the content, learning the structure, remembering the detailed phrasing, processing the information, and maybe modifying for future use. The various frequencies of which a song is composed will attenuate differently in the water; higher frequencies will travel less far than lower frequencies. Signals will sound quite different depending on whether the sender is near or far away. A whale would have immensely complicated computations to make in order to decipher the message as it was when sent. Clearly acoustic requirements in the sea could tie up a substantial part of the whale's brain capacity. (Bats, on the other hand, are involved in extremely complicated sound processing, using a brain of not more than one gram.) At the very least, study of the humpback whale song is going to give some clues as to the way whale minds are working.

Roger Payne's latest work is concerned with trying to relate songs to behaviour. Already the whale researchers have noticed that, during certain phrases of a song, the humpback will make particular movements of a flipper or the body. Underwater photographer Al Giddings recorded seeing a singing humpback 60 ft down moving its flippers back and forth in time with its song phrases. They seem to be dancing as they sing. Unfortunately, there are usually no other whales around to watch so perhaps they are merely flexing and stretching their muscles as they make the sounds.

Whale song was originally thought to be associated with courtship behaviour at the breeding grounds. Writing in *Marine Bio-acoustics* in 1964 William Schevill noted, 'The sonorous moans and screams associated with migrations of (humpback whales) past Bermuda and Hawaii may be audible manifestations of more fundamental urges...' Whale renditions have been compared with those of birds, crickets and gibbons in that they probably convey such information as species, sex, age, location, identity, readiness to mate, and readiness to engage in aggressive behaviour with rivals. Aggression is not a word often associated with whales but it is becoming clearer that whales are more aggressive than was first thought. Researchers from the University of Hawaii have described 'raw freshly bruised dorsal fins' and 'head nodules that appeared red and bruised'. Some have suggested that an assembly of singing humpbacks is reminiscent of a lek, where males gather on a communal courtship display ground to which females come in order to breed. It could be that females choose their mates on the basis of quality of singing, although there is no evidence of this as yet.

Behaviour associated with singing whales has been studied by Peter Tyack of Rockefeller University. In 1977, while recording humpbacks around Hawaii with Roger Payne, Tyack would go out in a small boat, looking at random for whales, put the hydrophone over the side and record the sound when he could hear one whale dominating.

> 'When you get into the water near a singing humpback whale it can be a scary experience. Your lungs resonate with the sound; it is very loud and you feel it throughout your body – you feel it more than hear – it is very eerie to be sitting 30 ft below the water just having your body reverberate with the sound of a whale. Occasionally when one of us is alone in a boat and has been out there for a few hours, he'll notice the whale next to the boat, ten or 15 ft away. The whale will often surface, come right up within inches of the boat itself, seeming to look the boat over, often staying for up to half an hour at a time, spy-hopping, lifting its head up to look into the boat, lifting its flippers up; it is very strange to have your wild animal come up to you when you don't expect it'.

Often Tyack was able to identify the singing whale because, when it surfaced, the sound level would be reduced. Occasionally, he noticed, the whale would stop singing and then surface with another whale, the two whales moving off together. Following the whales in a small boat revealed little of their behaviour, so at the end of the season Tyack was left with little more than a tantalising hunch that something interesting was going on. In the intervening months before the next season, Tyack and his co-workers devised an observational technique which was going to reveal fascinating information about humpback whale behaviour, in particular some insights into the function of song. The technique involved the combined operations of observers on a hill overlooking the whales, with others following individual whales in small boats (Boston whalers with outboards). A bay was chosen on the sheltered west coast of Maui, Hawaii, where large concentrations of whales appear each winter. In that way the overall pattern of movements could be seen from the hill while individual identities and vocalisations could be recorded in the boats. Each team was in radio contact with the other.

In the spring of 1979 they tried their first experiment. The first singer they followed stopped singing, joined with another whale and the two went off. During the season, 13 out of 28 whales that were followed after they had stopped singing joined with other whales. Indeed, the recording team learned to expect interesting things when a singing whale they were pursuing stopped singing. By observing whale behaviour in this way they were able to put together a picture of leviathan life below the sea.

A singing whale is almost always alone, separated by several hundred metres from any other whales. It moves slowly, turning this way and that while singing. It occasionally moves towards other whales but avoids any other whale that is singing. Singers do not appear to hold duets or to interact vocally in any way. They also do not seem to have strict territories. Singers are not found singing at the same 'song posts' on different days. In a few cases, Tyack observed singing whales approached by single non-singing whales. The original singer would stop singing, move along for a while with the intruder and then silently, but rapidly, swim away, seemingly displaced from his singing post. The new arrival would then start singing in the other's place. If a singer approaches a small group of whales such as cow and calf, he will often turn deliberately towards them. The cow and calf sometimes steer away, otherwise they continue on the same course, but they rarely move towards the approaching singer. With a two year breeding cycle, where a cow may conceive one year, have a calf the next, and then wait for a year until the calf is weaned before ovulating again, a cow may not be sexually receptive, and so will actively avoid a singing male. The singer often pursues the group and will attempt to catch up. If he is able to join them he immediately stops singing and becomes an escort to the cow. The cow, the calf and the ex-singer then swim along slowly and silently together, the calf keeping close to the cow. Escort whales were once known as 'aunties' and were thought to be female helpers. Sometimes they will interact with gentle flippering or rolling, behaviour previously observed in association with sexual activity in gray and right whales.

Normally any signs of aggressive behaviour are absent in cow-calf-and-escort groups. On one occasion, however, when Tyack was in the water with the whales, he somehow got between a cow and her calf, the calf having swum over to investigate Tyack. The mother, clearly upset, turned sharply towards Tyack and in so doing bumped into the surfacing escort whale which could not get out of the way in time. The escort produced a sound and the cow responded by emitting a series of grunts for about half a minute which might have been translated as 'watch out where you're going!'

Occasionally the cow and escort will dive below out of sight leaving the calf at the surface for a period of ten to 15 minutes, returning to the calf for five minutes and disappearing again. Whether mating takes place deep down in the ocean is not known, although researchers believe that this must be the case. Females ovulate at this time and males show an increase in testes weight (compared with the summer feeding condition) and increased sperm production. In addition females calve in winter and the gestation period is about one year. Songs sung towards the end of the season tend to be longer than those at the beginning. A similar observation in songbirds has been accounted for by an increasing concentration of testosterone in the blood. It is thought that similar events could be happening with humpbacks.

If the cow-calf-and-escort trio swim into the vicinity of another singing whale, the singer may begin to approach the group. Invariably the cow-calf-and-escort speed up and alter course away from the singer, but the singer seems highly motivated to join them. He will speed up dramatically, even while singing, and pursue the group, turning this way or that in an attempt to catch up. As soon as he has reached the cow-calf-and-escort he will also stop singing and a sudden change in behaviour takes place in the group. They begin to swim faster. The two escort whales engage in some kind of competition in which the secondary escort attempts to displace the principal escort from his position next to the cow. With the cow and calf up front and the two escorts behind the group may be swimming at ten to 15 km per hour.

Until now the original trio were silent but as soon as aggressive activity starts, a whole barrage of sounds can be heard. The whales will thrust at each other with their flukes or ram into each other as they jockey for position. The impact of fluke on blubber can be heard as a very loud slapping sound. Occasionally they blow bubbles underwater, an activity associated with aggression in many other species of whales. They also produce vocal sounds which the researchers recognise as social calls. There are many types of sounds – sometimes segments of songs out of context, high pure trumpeting calls and low grunts but as yet it is difficult to sort out their meanings.

The social sounds are loud and can be heard up to nine km away, thus advertising the location of the group. Other lone singing whales, on hearing the disturbance, will stream in to join the group. As more and more whales join, the activity increases. The rapidly enlarging group becomes very rowdy and engages in a great deal of aggressive behaviour. At the surface, flippers, flukes and heads can be seen being thrown out of the water. Below, the group appears to keep a particular structure. The cow and calf are the centre of activity with the cow as the nuclear animal and the principal escort alongside. The rest of the group take up stations around the pair and continually challenge the position of the principal escort by attempting to place themselves between him and the cow, the 'nuclear animal'. By now there might be up to 15 animals in the group.

Sometimes a secondary escort will displace the principal escort which will swim off. He may try to regain his position and is sometimes successful. However, the position of principal escort in a group is not long lasting. According to whale researcher Hal Whitehead of the University of Cambridge, who first described the 'nuclear animal-principal escort' structure, the principal escort retains his position for only seven or eight hours.

If one of the large groups has been swimming along together for a while, the activity seems to flag. Animals gradually lose interest and drop away. They take up well-spaced positions in the ocean and recommence singing. The cow-calf-and-escort group swims on.

In order to confirm some of these observations, Tyack and his colleagues set up play-back experiments. One immediate problem was to get sufficient volume of sound, and the help of the US Navy in the form of an enormous loudspeaker was the answer. In one test Tyack played back the social calls of an active group. The result was surprising and at times a little frightening. Singing whales would stop singing and rapidly home-in on the loudspeaker, charging the boat at high speed. Reaching underwater speeds of up to 12 km per hour, the whales would submerge and dive five metres from the boat. They would then circle for some time as if looking for the group of whales that should be making the sounds. When songs instead of social calls were played, virtually all whales moved away from the loudspeaker. Singers react to the playback song by terminating singing and swimming elsewhere.

It was difficult to follow these individuals without the song as a guide so it is unknown whether they are simply spacing themselves from the playback and then starting to sing again. Tyack, however, has suggested that one function of song is as a spacing mechanism for courting males.

The songs used in the playback experiments were always recorded in the same month as the experiments were carried out. Although it would have been interesting to see the reactions of the whales to an old song, Tyack and Payne felt that this kind of interference with a wild population might be considered a form of vandalism.

Still a complete mystery is the way in which singing whales produce the sound. It is presumably a vocal sound like our own but this would involve air and bubbles are not obvious during singing. Inside the whale's head there is a great deal of complicated plumbing. It could be that air is shunted back and

Fig. 9 Humpback whale cow-calf-escort group. Fig. 10 (*opposite*) Humpback whale: cow-calf-escort groups: A. Cow and calf are joined by an individual that has stopped singing. B. Cow, calf and primary escort. C. Cow and primary escort dive below, leaving calf at surface. D. Cow, calf and primary escort. E. Cow-calf-escort group attract secondary escorts. F. Secondary escorts fight for primary escort position. G. Unsuccessful escorts leave group and recommence singing. H. Primary escort leaves cow and calf and recommences singing.

forth across the vocal chords in these tubes and cavities. Singers perform their underwater arias at depths of 80 to 100 ft below the surface. A deep submarine canyon will reflect the sound so that it reverberates as if in a gigantic underwater cathedral. A shallow sea bottom produces almost studio-like recordings. Whales stay singing below for up to half an hour at a time, returning periodically to the surface to breathe. As they approach the surface the sound gets quite a bit fainter. Whether this is due to the whale actually singing more quietly or caused by some acoustic property of seawater is not known. When breaking the surface they do not stop singing or break the rhythm of the song, but during four or five identifiable pauses in the song they will take breaths before diving below once more. There is a particular song theme in which breathing tends to occur so Tyack and his colleagues were able to anticipate a singer's arrival at the surface. As they tip their flukes and sound, the volume of the song becomes much louder, literally vibrating through the boat.

In the feeding grounds in polar regions the whales are relatively quiet. They do, however, occasionally indulge in social conversations using sounds similar

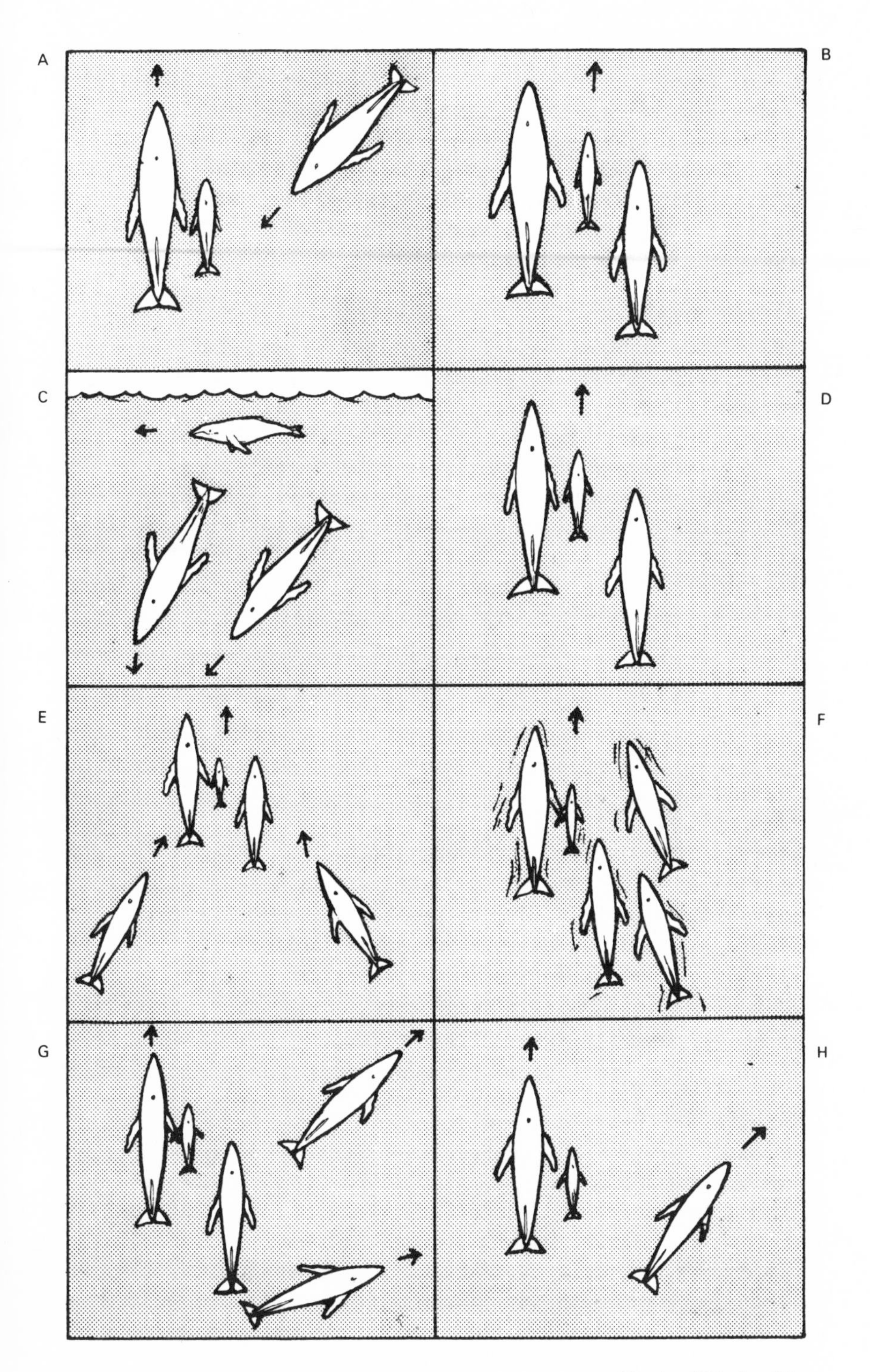
A
B
C
D
E
F
G
H

to those emitted by the large rowdy groups. During a preliminary study in the Glacier Bay area of Alaska, Tyack observed that the sounds are usually made when the whales are in groups, particularly when individuals meet and join or split up. Bill Dolphin, at Boston University, has made a more extensive study but is not sure whether the sounds are greetings or keep away calls. Whales frequently feed in organised groups, sometimes in line or chevron formations, so competition for resources is unlikely to be the reason for any spacing calls. Social calls could be used to keep a feeding group together but there are no observations of this as yet.

Humpbacks have one unusual form of feeding in which sounds are often heard, called bubble-net feeding. Starting low in the water, the whale will lay a circular column or net of bubbles in the water. As the circle is closed the whale will come up right in the middle of the bubble tube and engulf the food in the centre. Tests on the density of food in the bubble-net have shown it to be up to 50 times higher inside the circle than in the surrounding water. Pleats in the lower jaw swell out with each mouthful. The whale's jaw is then raised and the pleats flatten forcing the water out between the baleen plates and leaving a concentrate of krill and small fish which can be swallowed. According to Charles and Virginia Jurasz from Glacier Bay, Alaska, the diameter of the bubble is selected and nets with a variety of mesh sizes can be made.

Southern Right Whales

While Peter Tyack and his colleagues studied humpbacks off Hawaii, another group of Rockefeller University whale-watchers, working with Roger Payne, directed their attentions to another great gathering of whales, this time southern right whales *Eubalaena australis*. Reaching lengths of 50–60 ft, these enormous blue-backed whales are characterised by the 'bonnet', a collection of white, crusty patches on the top of the head and on the snout which often provide anchorage for whale lice, barnacles and sea anemones. These callosities, as they are known, probably serve similar functions to human eyebrows and other facial hair, in the case of the whale deflecting water from entering the blow-hole when at the surface. Those on the snout seem to function as 'antlers' during aggressive bouts.

The centre of activity was the Peninsula Valdes in Patagonia – an enormous cape enclosing two shallow, almost land-locked and very desirable bays; desirable for whales that is, for Peninsula Valdes is renowned for its high winds and rough seas. The windier it is the better southern right whales like it. As the wind increases and white-horses appear on the sea's surface, the whales burst into life and literally play in the storm. An individual will leap from the water, crashing back in an explosion of spray. Sometimes a huge tail is seen sticking vertically out of the water. Observations have revealed that southern right whales actually sail in the wind. High winds are associated with much activity including lob-tailing (slapping the water with the tail fluke), flipper-slapping and breaching. During a storm, underwater noise increases, particularly as a result of waves crashing on the shore, thus masking the lower frequencies with which southern right whales communicate. By slapping the surface, individuals can keep in touch.

Investigating sound communication and its importance in the social behaviour of southern right whales are Rockefeller University's Christopher and Jane Clark. Originally a biochemical engineer, Chris Clark got involved by accident – he let Roger Payne borrow his pick-up truck, and after a few visits to the Payne household became hooked on whales. Roger Payne and his family had already spent five seasons recording southern right whales at Peninsula Valdes, and their enthusiasm was clearly infectious. Chris Clarke's expertise in electrical engineering was to help solve the problem of following whales under water. Using an underwater hydrophone array linked to a portable mini-computer, Clark and his colleagues were able to observe the whales' movements and interactions from the cliff-top while at the same time plotting the positions of the whales making sounds.

The most prolific sound to be heard in the bay is a simple low frequency, tonal call which is thought to be used for contact over long distances. As individuals approach one another they exchange an increasingly rapid series of calls, and having met stop calling altogether. These calls can be heard at any time of the day during the five months of the year in which whales congregate in the bays. Contact calls have a frequency between 100 and 200 Hz, which happens to coincide with the quietest range of frequencies in the usually noisy bay. Southern right whales in the bay appear to be using a low-noise sound window through which to communicate and so make contact with others over great distances with the minimum of effort. Whales calling in the bay, a shallow cathedral-like dish 25 kilometres long by 15 kilometres wide, can be heard plainly from one side to the other. In deeper water the sounds should travel even farther.

Often, several whales will come together to form a tight active group. Five 40-ton males and three females, for example, might be swimming in an area the size of a gymnasium. In these active groups the sounds tend to become more excited, rising in pitch, with mixtures of high melodic notes and low, pulsed growls – the higher the pitch, the higher the excitement. It is difficult underwater to work out which whale is making which noise, but it is already clear that when a large number of males get together aggressive growling sounds dominate proceedings. It has been suggested that they are competing for the female, prior to mating. If two males are left with a female, bouts of growling are heard. Eventually one male departs and the sounds change to more high-pitched melodic notes. Chris Clark describes the vocabulary as a continuum, rather than a set of discrete sounds except, that is, for the stereotyped contact call. Curiously southern right whales don't have any alarm calls, although frequently harassed by killer whales. Indeed, the calls seem not to contain any complicated messages. Right whales are grazers, and therefore do not require a sophisticated communication system to coordinate activities such as hunting. They are promiscuous rather than monogamous, so few social sounds are required. Their only need for calls is to bring the 'herd' together.

Once he had identified the southern right whale's repertoire and suggested possible functions for the sounds, Chris Clark used playback experiments to determine whether the interpretations put on the calls were indeed correct. His first task was to see if whales responded to contact calls. A variety of sounds was collected for the experiment. In addition to the right whale contact calls themselves, Clark tried background noise, pure tones at frequencies contained in whale calls (i.e. 200 Hz), humpback whale songs, and even Handel's Water Music! An underwater loudspeaker was placed on the bottom of the bay, not far from the hydrophone array, in front of the observation hut. To make identification easier, the observers waited until there were only two or three whales in the area. A whale would be spotted by the pattern of callosities on its head and its calls recorded for a quarter-of-an-hour before the playback experiment. Playbacks were started when the whale had passed the loudspeaker and hydrophone array, and was heading out to sea. The observers noted any sounds the whale made in response to the playback, and its direction and speed. The whales responded only to playbacks of southern right whale calls. They would stop, turn around dramatically in a flurry of foam, and swim at full speed back to the loudspeaker, at the same time producing more sounds. When other playbacks were used, whales would ignore them and continue to swim away. On one occasion, at the end of the season, when few whales were in the bay, a lone right whale was exposed to playback of a group of whales. It responded by returning to the speaker. No more sounds were played and the animal began to swim off. After a little while it turned back and swam to the area of the speaker. It continued to do this until dark when Clark left it alone in the bay silently searching for its fellows.

Having shown that southern right whales can differentiate their own calls from other similar sounds, the Clarks hope to continue playback experiments in order to determine the biological functions of the sounds in a whale's acoustic repertoire.

Scattered Herds

For many years a mysterious low frequency sound has been heard in the oceans of the world. The sound is pulsed in 'blips' with a frequency close to 20 Hz. With a bandwidth of only 3 Hz, the '20 Hz signal', as it is known, is almost pure tone. It is also of high intensity, so loud that scientists at first believed it could only come from a non-biological source such as surf on a distant beach. The sounds, though, increase in late afternoon, reaching a peak around midnight, and waves on beaches don't usually have such a pronounced diurnal rhythm.

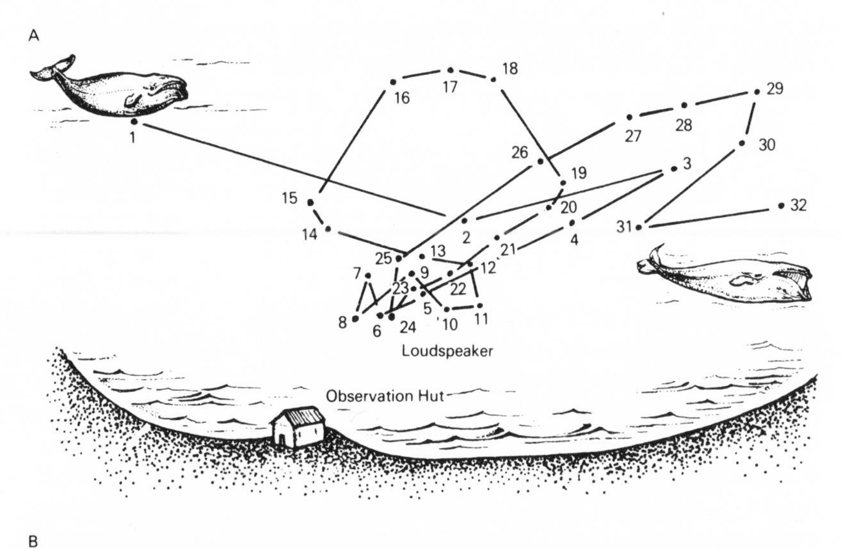

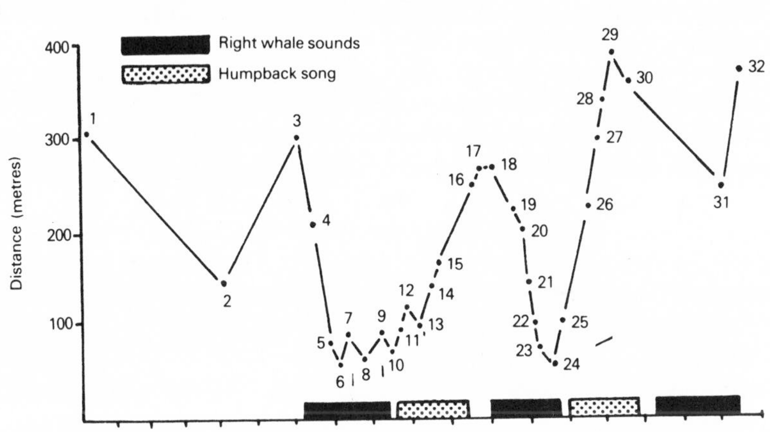

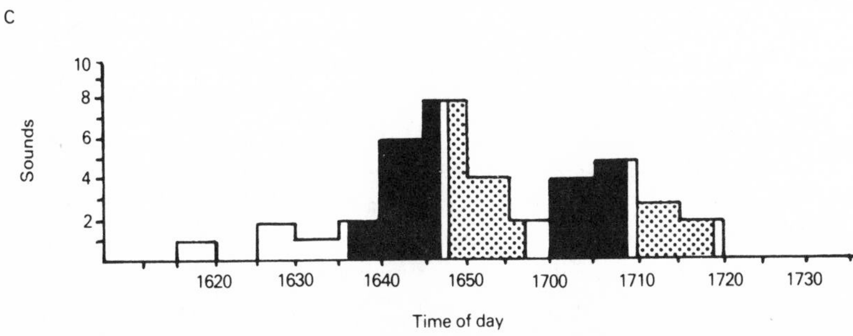

Fig. 11 Response of Southern right whale to the playback of right and humpback whale sounds: A. The path of the whale. B. The distance of the whale from the loudspeaker. C. The number of sounds made by the whale per 5-minute interval.

Listening to underwater sounds off Bermuda, B. Patterson and G. R. Hamilton noted the spacing of the '20 Hz signals'. Trains of pulses were heard for about 15 minutes, separated by two-and-a-half minutes of silence, a pattern reminiscent of the breathing cycle of a large slow swimming whale.

Using an array of hydrophones, William Schevill and his colleagues from Woods Hole Oceanographic Institution, Massachusetts, were able to home in on the sources of the '20 Hz signal' and each time it turned out to be coming from a fin whale *Balaenoptera physalus*, the second largest animal ever known to have lived.

But why should fin whales want to make these very loud low frequency sounds? Most whales are thought to be social animals. The smaller toothed whales, like dolphins and killer whales, group together in schools, pods or herds, sometimes in large numbers. The baleen whales, on the other hand, are only found travelling alone or in small bands of no more than 20 animals. Could it be, though, that the great whales are indeed in herds, but in herds many hundreds of miles across? Roger Payne and Douglas Webb at Woods Hole believe there could be something in this hypothesis. They believe that fin whales might be in contact with each other using these low frequency sounds. They do not suggest that meaningful and complex messages are winging their way across the ocean, but simply that acoustic signalling might be used by whales to locate one another, to aid future rendezvous, and to keep them together in their widely dispersed groups. A lone whale, therefore, just might have company, albeit a hundred miles away over the horizon.

Two fin whales, several kilometres apart, emitting individually pulsed 20 Hz signals were recorded by Patterson and Hamilton off Bermuda. The researchers were able to follow the direction the whales were heading with the aid of a multiple hydrophone array. The first whale called for about three hours while going south. The second whale, about five kilometres to the east, then began to call and the first whale changed direction towards it. Was the first whale being guided to the second by sound? Follow-up work in this area is prohibitively expensive so as yet there is no answer, but the simple and repeated patterns of pulsed sounds *are* ideal for long-range communication. Fin whales, unlike humpbacks and grays, do not appear to breed always in the same areas, so there must be a mechanism to bring individuals together. The 20 Hz signal seems to fit the bill. It is lower in frequency than the noise generated during turbulent storms; loses little energy when bounced off the sea bottom; and is apparently the best frequency for propagating through a surface-frozen polar sea. But the ocean is a noisy place. The whales would have to make themselves heard over considerable background chatter, particularly in today's ocean where man's supertankers and other technological advances pollute the ocean not only with oil but also with noise. How do they do it?

There is one published case of a small underwater explosion from just four pounds of dynamite, detonated off Australia and detected at Bermuda about

12,000 miles away. That *is* unusual, but calculations made by Payne and Webb indicate that fin whales could communicate over considerable distances by making use of deep water sound channels. These acoustic channels are the result of physical characteristics of the ocean. Differences in water densities, salinity, temperature and ocean currents, produce a channel at a particular depth which tends to trap sound. It works as a kind of underwater voice-tube by concentrating sound energy in a narrow beam rather than diffusing it over a large area.

Without the help of a deep ocean sound channel, individuals could speak with one another over a distance of about 50 miles. Making use of cylindrical propagation in a deepwater sound channel, the distances could reach a maximum of about 500 miles. In pre-steamship days, these distances would have been further increased to 140 miles and 3,500 miles respectively, and these are conservative estimates. Clearly the fin whale acoustic system must have evolved in a quieter ocean and it is conceivable that man's maritime developments could have seriously impaired communication between whales and upset their lifestyle – a further setback following their wholesale slaughter at the beginning of this century. So far, researchers have not demonstrated that whales actively seek out the sound channel, but inevitably their calls will spill into the channel and the sounds propagate over long distances, audible to any whales listening at channel depths.

The sensitivity of hearing in baleen whales is thought to be excellent, albeit in the lower frequencies. Fin and humpback whales have many more fibres in the nerves from the ear to the brain than either man or the bottlenose dolphin. They appear to lack the high-frequency or ultrasonic hearing capabilities of dolphins but they may hear very low sounds in the infrasonic frequencies. The blue whale *Balaenoptera musculus*, for example, the largest creature ever known to have lived on earth, is thought to emit very low frequency grunts, but surprisingly there has been the suggestionn that it is also capable of vocalising in ultrasonic frequencies. The minke whale *B. acutorostratus*, one of the smallest of the baleen whales, calls with a very low note. The Californian gray whale *Eschrictius robustus* makes moans, bubble sounds, knocks and 'a metallic-sounding pulsed signal'. They also produce clicks which are probably used for navigating and locating food.

The sperm whale *Physeter catodon*, a large odontocete or toothed whale, produces clicks which are thought to travel long distances under the sea. William Watkins and William Schevill at Woods Hole have been analysing the sounds made by sperm whales. They have a multiple array of hydrophones and can locate accurately the positions of individuals.

Groups of these large, toothed whales are heard to make clicking sounds and each whale produces its own distinctive pattern of clicks like a morse code. Using this unique identification and location technique, Watkins and Schevill have been able to track individual sperm whales and observe their behaviour. They have found, for example, that sperm whales surfacing within

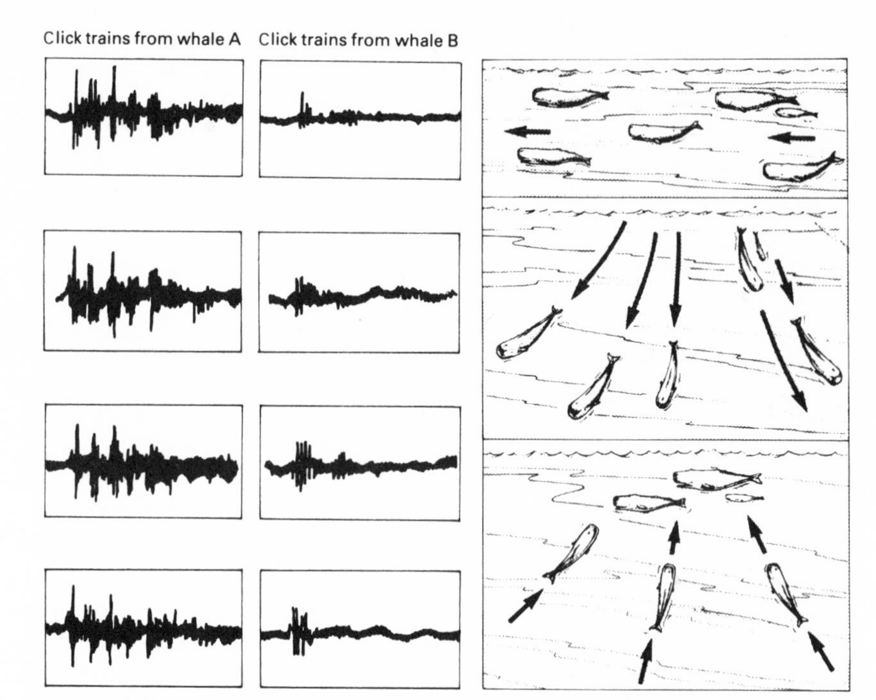

Fig. 12 Sperm whale herd on the move, each individual producing its own sound signature of clicks.

ten metres or so of each other spread out like an inverted funnel when they dive again, and so are separated by much greater distances as they reach the bottom. Returning to the surface they emit more clicks and gather together once again in a close-knit group. Sperm whales are more obviously gregarious than some of the other large whales, and their click signatures would be important in a complex social structure where individuals might want to contact each other during mating, feeding and so on. Groups of sperm whales are thought to be led by a dominant animal which would use sound to guide its family group. Whalers were aware of messages passing between sperm whales. They were sure that distress or alarm calls were given by harpooned animals which seemed to elicit responses from others many kilometres away.

Although we still know surprisingly little about the behaviour of humpback, southern right, sperm, fin, blue and minke whales, we do know that, despite international agreements, they are still threatened by the activities of our own species. By the time we understand sufficiently about them to begin the urgent work of conservation, it may be too late. The damage may be irreparable. In the future their beautiful calls might exist only on long-playing records and the research tapes of scientists.

3

GENTLE KILLERS

Dolphin Language

Killer whales, pilot and sperm whales, dolphins and porpoises are all classified together as the odontocetes – the toothed whales. Characteristically they are all noisy animals, but the voice of the dolphin is perhaps the most familiar. All over the world, 'singing dolphins' have thrilled and amused audiences in dolphinaria and marine circuses. As far back as the 4th Century BC Aristotle wrote of dolphin squeaks and moans, but in AD 1983 we still don't know exactly what they mean, or indeed even how they make their sounds. It is probably the mystery and apparent charm of these animals that has most contributed to the stories about their abilities. The almost complete lack of understanding of their behaviour, coupled with their perpetually smiling faces, makes them irresistible as objects of affection. It is perhaps because of this close affinity between man and dolphin that dolphin behaviour is often interpreted in the same way as human behaviour. According to some researchers, the mythology obscures the science.

The early Doolittle obsession with wanting to talk to dolphins stems from the fact that dolphins have large brains, as large as those of humans. Surely with a brain that size, so the theory goes, a dolphin has something worthwhile to say? This is further supported by the revelation that the neo-cortex – the part of the brain with which we create, innovate and reason – covers 98% of the surface of the dolphin's cortex. In man the figure appears to be 96% and, just for comparison, in the unfortunate kangaroo a meagre 69%. Unfortunately, anatomy also reveals that the dolphin's neo-cortex is much thinner than in humans, so this line of reasoning is rather inconclusive.

It was thought at one time that, contained within the dolphin's wide range of vocalisations, were complicated messages that could rival the complexity of human speech. Dolphins were thought to have mystical, paranormal powers and were attributed with super-intelligence. They were thought to be among the cleverest animals on this planet. Doyen of the 'smart dolphin movement' is John Lilly, a doctor who has studied such topics as neurophysiology and hallucinogenic drugs. His object has been to discover a means of communication between dolphin and man in order to understand dolphin language, culture, philosophy and even their system of ethics.

In an early experiment, Lilly had dolphins mimic English. They could follow a count up to ten, and almost say some simple English words. The recordings are amusing, but mynah birds and budgerigars can do better. Lilly's early mimicry experiments, however, did turn up some interesting information. Dolphins, it seems, showed a remarkable ability for rapid mimicry. On some occasions an animal would begin mimicking before the signal it was copying was finished. The time domain, therefore, was quite different from that in budgies. In addition, in another experiment, a researcher would read out a series of syllables, consisting of vowel and consonant sounds, and the dolphin would mimic the same number of sounds with the same durations with 70% accuracy. Dolphins obviously appreciate quantitative aspects of human speech much better than do mynahs and parrots. English, though, was clearly not the language with which to investigate the linguistic abilities of these remarkable creatures. Today, teams from California, Florida, Hawaii and The Netherlands are carrying out numerous costly experiments using signs, symbols and computer-assisted sound equipment, all witness to man's need to communicate with another animal.

There is, of course, a serious side to this kind of research. By 'conversing' with the study animal, information can be sought about its cognitive characteristics and an understanding gained, perhaps, about its intellectual abilities and limitations. Lilly himself is involved with the JANUS (Joint Analog Numeric Understanding System) project in California. A sophisticated computer-assisted sound system receives and transmits bleeps and whistles to and fro between captive dolphins and their human researchers. Dolphin sounds are matched to computer generated sounds, and to visual letters and other symbols which the dolphins can see on an underwater screen at the side of their tank. The dolphins react to the symbols on the screen, cause the sounds or symbols to change or simply match the sounds with dolphin sounds. The aim of the experiment is to create a new form of 'language' that is mutually accessible to both man and dolphin. Tests have started with 48 or so sound and symbol combinations, known as morphemes. Each sound-symbol combination is associated with an object, a place or an action. Lilly is using the computer interface to link the dolphin's high frequency (2,000–4,000 Hz) vocalisations with the human's relatively low frequency (200–2,000 Hz) speech and to change the time domain so that the dolphin's rapid utterances can be directly translated into the relatively slow delivery of human speech. Lilly estimates it will take about five years to work out a human-dolphin dictionary which could be used to communicate across the species boundary.

Flipper Sea School in Florida, once host to the famous television dolphins of the same name, has adopted a similar approach to training in order to explore a dolphin-human communication system. A whistle language, similar to that used to train sheep dogs, is being used to instruct the dolphins. In addition, one dolphin is responding to human words as well. At the Dolphinarium, Hardervijk, The Netherlands, Wilhelm dodoc Van Heel has been working with a killer whale called Gudren. Frequency modulated tones associated with objects in the tank were played to the killer whale. When the signal was given Gudren was expected to touch the relevant object. This she did. If the wrong object was presented with a particular sound signal the whale became very upset. Eventually Gudren was able to imitate the object-linked signals and touch the correct object. Van Heel then introduced sound signals, to represent action words, verbs, and the whale went on to recover objects on sound cues. The remarkable thing about this particular whale, though, according to the television programme *Talking Whales*, was that Gudren began to talk back. During a training session she was heard to give the sound signal to fetch the object in the tank, a copy of the signal the researchers had given her. This was the first time the conversation had been two-way.

In Hawaii, Dr Louis Herman, director of the Marine Mammal Research Laboratory of the University of Hawaii, together with Dr Douglas Richards and Dr James Wolz, has been working on another major project with two female bottlenose dolphins *Tursiops truncatus*. The dolphins have learned to respond to visual and verbal signals in much the same way as a sheep dog; but there the similarity ends, for the dolphins talk back. The background to Louis Herman's current research interests includes many years of observations on the dolphin's sensory systems, particularly its capacity to see and to hear, and studies of its learning and memory abilities. The dolphin, for example, has a good auditory memory, being able to remember long lists of sounds presented to it. Herman therefore wanted to know if language ('the most complex method known of information transfer from one creature to another') could be added to the growing list of dolphins' abilities. Instead of looking at the way a language might be produced by a dolphin, the research team concentrated on how an animal might understand language. To do this, Herman copied the technique used by second-language teachers who instruct pupils by getting them to carry out tasks. The level of understanding can be measured by how well the task is carried out.

This involved an entirely new method of training. At a dolphinarium the normal way to get an animal to learn a new trick is to entice it gradually with food. To teach a dolphin to go through a hoop, for example, the animal is presented with a hoop, rewarded initially for poking its head through, and then rewarded progressively as more and more of its body passes through until it swims through completely. A gesture or sound signal would then be provided to initiate the entire behaviour of swimming through the hoop. Herman's technique differs in that the dolphin is first taught the word for 'hoop' and then the word for 'through'. The words can be combined into 'through hoop' and without special training the dolphin will know what to do. If a new word such as 'gate' is introduced and linked to 'through' the dolphin knows to go through the gate as it did the hoop. An object word and an action word are taught to the dolphin and combined with other words in new contexts; this gives the researchers flexibility in the training system.

Unlike sign-language experiments with primates, where chimps and gorillas are invited to converse with the trainer, thereby producing 'words' which might be open to misinterpretation Herman is simply examining the cognitive characteristics of dolphins with what he feels are more objective interpretations of the results. Herman is simply asking questions. In what do dolphins specialise? What are their limitations? What are they good at? How do their successes and failures compare with those of other animals? What role does intellect play in a dolphin's world? In short, Herman's main object appears not to be language but to create a tool for discovering a whole range of behavioural abilities in dolphins.

Dolphin Sounds

What of the dolphins themselves; what are *they* saying to each other? Probably the most famous experiment in dolphin communication research was carried out in 1965 by Jarvis Bastian of the University of California at Davis. He placed a pair of bottlenose dolphins in adjacent tanks so that they were isolated visually, yet could still hear one another. The female was taught to push paddles in order to receive a reward. The male was offered another set of identical paddles but received no training. He learned, however, to push the right paddle. The only way the male could have obtained the information would have been from the female next door. Throughout the tests both animals were heard to emit many whistles, squeaks and clicks. Although the information must have come from the female, unfortunately the results were not conclusive proof that deliberate communication had taken place. The male may have picked up sounds inadvertently made by the female and quite independently trained himself to use these to his own advantage; a remarkable feat in itself.

There have been many stories of dolphins in the wild using sophisticated communication systems. It is reported that dolphin schools will avoid boats out to capture them. It has been recorded even that dolphins will steer clear of a particular type of boat simply because the shape is similar to those that are out to do them harm. It was thought that somehow or other dolphins were able to tell each other 'to avoid that boat over there; one of the school was killed by a similar-shaped boat last week'. On the other hand, dolphin schools are hauled out and killed in their hundreds by tuna fishermen around the shores of Japan. No complicated communication system saves them. If messages carrying sophisticated information are passing to and fro, why don't they avoid the trap? Indeed, why do so many small whales get themselves beached to die of exposure in the sun? The paradox remains.

What *is* clear is that dolphins have a varied vocal repertoire. There are squawks, whistles, squeaks, burps, groans, clicks, barks, rattles, chirps and moans. The bewildering array of sounds can roughly be divided into two types – pulsed and unpulsed sounds. The pulsed sounds include clicks and burst pulses. The bursts may be arranged into chuckles, chirps and click-trains. Trains of clicks sound like rusty hinges being opened, or the whine of machinery, while some burst pulses have been variously described as 'raspberries' or 'Bronx cheers'.

The more continuous sounds are the whistles and squeaks of frequency-modulated pure tones which may last for several seconds. It was thought by early researchers that the whistles are the main communication sounds. This may have been because the whistles are easier for humans to hear and to study, the clicks being mostly in the ultrasonic frequencies. But this was perhaps yet another piece of scientific mythology associated with dolphins and was brought into doubt when it was revealed that many odontocetes, including river dolphins, sperm whales and many other openwater dolphins, have not been heard to use whistles; they only emit pulsed sounds.

Many of the pulsed click-sounds are used for echolocation and navigation; some, though, are thought to have social functions. Male bottlenose dolphins and harbour porpoises *Phocoena phocoena* have been known to give pulsed 'yelps' during courtship. Similarly Atlantic spotted dolphins *Stenella plagiodon* 'squeak' during training sessions in captivity. Frightened or distressed dolphins emit pulsed 'squeaks', which could be a type of alarm call. Aggressive 'buzzing' trains of clicks are heard when two males confront one another. What seem like exchanges of burst-pulses have been recorded between individuals in a school of Hawaiian spinner dolphins *Stenella longirostris*; so, too, between pilot whales *Globicephala melaena*, and between narwhals *Monodon monoceros*. All three species are known to whistle.

As yet, the social function of click-sounds has been little studied. There are, however, a few observations which allow some generalisations to be made about those animals that use whistles and clicks and those that use just clicks. Heaviside's dolphin *Cephalorhynchus heavisidii*, the harbour porpoise, the finless porpoise *Neophocaena phocaenoides*, and the pigmy sperm whale *Kogia breviceps*, have pulsed sounds only and are all odontocetes that aggregate into relatively small groups of three to 20 individuals. Hawaiian spinner dolphins and bottlenose dolphins, on the other hand, often form huge schools of hundreds of animals that have a tendency to forage together. They are the whistlers. The very high frequency clicks travel less far in the water than the relatively lower frequency whistles, and would be more suitable for sending messages between individuals of a small group. Whistles would carry better between dolphins in a large and spread-out group.

There are suggestions that emotional meanings can be read into these sounds. Abrupt, loud sounds, for example, are given in aggressive situations. Sometimes these vocalisations are accompanied by non-vocal 'jaw-clapping' or 'tail-slapping'. Intimate chuckling sounds are heard during bouts of caressing and touching. Dolphins touching, for example, make a sound much like the noise you get when rubbing a finger against a wet balloon. There are signature whistles. A dolphin will produce a whistle that seems to be its own whistle, which appears to be used as a long-range individual-identifier-signal across the width of the school, to let every dolphin know where all the others are placed.

In an experiment in 1974, in the USSR, two neighbouring bottlenose dolphins, linked by sound only, both produced whistles and rhythmically related clicks. It was suggested that dolphins, like songbirds, have an opening identification portion of the call, the whistle, followed by a more complex message portion, the clicks. This birdsong parallel was continued further in the interpretation of the results of an experiment with tropical spotted dolphins *Stenella attenuata.* A male was captured and recorded. It produced almost continuous bouts of whistling which were probably alarm or distress calls. The calls were played back to the school from which it came and they fled, instantly. The same calls played to another school elicited curiosity and not flight. The captured dolphin's own school detected danger from the familiar, but frantic, call of the subject, whereas the 'strangers' were unable to appreciate the significance of the calls. Does each school, then, have its own vocabulary of calls, maybe as a result of mimicry within the school? Dolphins are good mimics, as witnessed in early communication experiments, and related species, such as killer whales, are known to have distinct local dialects.

Killer Whales

The evidence of dialects among killer whales *Orcinus orca* was demonstrated by Dean Fisher and John Ford at the University of British Columbia in Vancouver. One of their first discoveries was that the sounds of any given killer whale pod are very stable. Unlike humpback whales, killer whales do not change their songs through time. In a variety of situations the animals make the same associated sounds over and over again. In several different pods, John Ford has been able to identify, on average, 12 distinct stereotyped or discrete calls which are exchanged between whales when spread out and foraging maybe over an area of a couple of miles. It is thought that these calls are emitted in order that individuals in the pod can keep in touch with each other, although out of direct visual contact. It is not known yet whether each of the 12 sounds has a different meaning because, below the sea, it is difficult to link sounds with any particular pattern of behaviour. The Vancouver research team feel that the most likely information being communicated includes their position, individual identity, emotional or activity state, and pod identity through the dialect. It looks unlikely that a more highly structured and sophisticated message, like human speech, is being exchanged. One call is given, the others in the group respond, and then they all switch to another type of call.

Dialects have arisen as a result of the isolation of one pod from another. Killer whale pods are essentially extended family groups. Once an animal is born into a pod it rarely leaves the group. In this way a pod may grow up to 50 strong, although the average family group consists of between six and 15 individuals. If an ancestral pod grows to a certain size, it is likely (although not proved) to split into two or more smaller groups, which spend progressively less time together. Initially the calls given by the groups would be similar but as time went by, probably over several decades, the dialect of each group would drift away into its own distinct sound and shape. Around Vancouver, for example, resident pods hunting in the same bays and having their own clear patrol areas tend to have very similar calls, whereas visiting migratory or transient groups can be heard to have distinct calls indicating a quite separate ancestry.

Hawaiian Spinner Dolphins

One group of dolphins whose social interactions have been studied fairly extensively are the Hawaiian spinner dolphins of the Pacific. Much of the early work was with dolphins in captivity, but increasingly today researchers prefer to work with natural, relatively undisturbed groups of animals in the wild. Dolphin watchers, Professor Ken Norris and Sharron Brownley, from the University of California at Santa Cruz, have been observing dolphins in the wild, in particular the Hawaiian spinner dolphin – easily identified by its twisting leap. They have been looking at the structure of dolphin schools, at where individuals move, how they interact socially, and listening for the sounds they make during different behaviour periods throughout the day and night. Spinner dolphins, according to Sharron Brownley, are very convenient animals to study for they do everything together. They sleep, play, feed and travel together, and over the course of the day act in a predictable manner. At dawn a large group might arrive from night-time feeding and split into smaller groups as it gets close to the shore. In bays around the Hawaiian islands they spend the day resting, playing and generally socialising. In the late afternoon activity increases with what is known as zig-zag swimming. Individuals swim back and forth until the entire group move once more out to sea in order to feed during the night.

Throughout these activity periods the spinner dolphins are vocal, with different sounds and sound patterns at different times of the day. When they are resting they are quiet. They swim slowly, close together, and make few sounds, usually a few clicks. Gradually, during the afternoon excitement rises and they begin to increase the number of whistles and burst-pulses. The whistles are thought to represent individual identities, while the burst-pulse signals indicate to each other their emotional states, whether angry or playful. Each activity period blends with the next. The sounds then change from predominantly burst-pulses and few whistles to more whistles and fewer burst-pulses. On the surface, activity hots up considerably as dolphins are seen leaping and twisting out of the water. Underwater they roll over each other and play, all the while emitting sounds. The more physical activity, the more noise. Sharron Brownley feels that when the animals go into zig-zag swimming it is almost as if they are trying to decide whether they are ready to go to sea to hunt. They swim slowly to the entrance of the bay and then rapidly swim back. Cape hunting dogs in Africa go through a similar ritual before hunting. Here cooperation is the key to successful hunting. Before leaving for the plains they go around twittering to each other, getting progressively more excited about the hunt and therefore more ready to hunt as a unit. The pitch of excitement rises and the level of noise rises until all of a sudden they all take off and track down their evening meal. Similar kinds of prehunting calls are heard from other animals that hunt in groups, such as wolves and hyaenas. Here a dominant animal assembles the group. In wild Hawaiian spinner dolphin schools it is difficult to identify whether an animal is male or female, let alone dominant or submissive, although researchers believe there may be group elders which shape the direction and guide the school. In other dolphin species, such as bottlenose dolphins, dominance hierarchies *have* been observed where a large female is dominant over other females and sub-adults and a large male is dominant over the entire school.

In Hawaiian spinner dolphins, however, when one dolphin starts calling, the rest tend to chip in too so there are periods of hectic vocalisation followed by periods of silence. Sharron Brownley suggests that the bursts of noise may indicate how unified the group feels. When all the dolphins chime in at the proper time they know they are all ready to hunt. If some do not join the chorus then it might indicate they are not ready to go. The chorus then starts again until everybody is alert and ready.

The peak period of vocal activity, then, is late afternoon. The noise level rises noticeably. The wild cacophony of sound is so incomprehensible and so full of all kinds of sounds that Ken Norris has nicknamed it the 'Yugoslavian-news-report'. Non-vocal sounds also accompany the activity. Animals leap from the water, slapping their heads and tails, and generating loud percussive noises which may help whip up excitement in the school. The leaping and spinning seems to be infectious, spreading rapidly through the group. Both visually and acoustically, leaping, slapping and whistling serve to tell where each is located and whether ready to go. Once at sea they use clicks for navigation and hunting.

Dolphin Whistles

Whistling associated with feeding has been observed with many different odontocete schools, both wild and captive. In dolphinaria bouts of whistling coincide with the main feeding periods. Douglas Richards at the University of Hawaii made recordings of a pair of newly arrived bottlenose dolphins and found that they whistled least at night (the period of low activity for this species) and most early in the day. A year later whistling became synchronised with routine feeding and training sessions.

Dolphins riding the bow-wave of large boats or humpback whales are heard to whistle, as are individuals in unfamiliar situations, such as those stranded, captive, or otherwise isolated from the school. The level of excitement is accompanied by a comparable level of whistling. Encounters with unusual objects or potential predators, on the other hand, will elicit silence or reduced vocal activity. When two schools meet, whistling activity may increase or decrease dramatically. It is not clear why. Animals harpooned whistle continuously, as will mothers separated from babies. Captive dolphins whistle when introduced to their new tank, although they quickly settle down.

Many attempts have been made to catalogue dolphin whistles. Often, several identifiable whistles are recorded together with a multitude of minor variations. It has been suggested that these animals have a graded system of vocalisations, where each sound type overlaps with another to form a series of sounds rather than discrete signals. The same kind of graded series has been described for certain monkeys and apes .

Dolphin Clicks and Echolocation

The other main category of dolphin vocalisation, pulsed sound, is often, and in most cases mainly, used for echolocation and navigation. Echolocation is sometimes considered as autocommunication or communication with self, and broadly meets the definition, used earlier, that an animal has communicated when it has transmitted information that influences a listener's behaviour. Most of the toothed whales are thought to be able to interrogate their environment with sound. Using echolocation clicks a dolphin can see with sound. By bouncing sounds off an underwater target and analysing the signal it gets back, a dolphin is able accurately to locate the object, determine whether and where it is moving, discriminate different object densities, say fat from bone, can tell whether a target is dead or alive, and if alive and potential food may be able to stun it, and sometimes kill it, with a high-intensity beam of sound. That dolphins are able to echolocate was revealed in 1942, and although since then experiments have mushroomed, we still know very little of the nature and function of a dolphin's echolocation system.

One of the workers who has been at the centre of dolphin research, right from the early days, is Ken Norris of the University of California at Santa Cruz. His pioneering work was with captive animals in dolphinaria or marine circuses; indeed, the only way to fund serious research was to teach dolphins new tricks. One of his first tests was for a TV show. Ken Norris was allowed the use of a dolphin on the understanding that it could be televised, and therefore also it had to be entertaining. The researchers noticed that when a dolphin swam between a pair of hydrophones placed in the tank the sounds would increase in volume if it swam directly towards one hydrophone. Then as it turned and swam towards the other, the sounds would appear to come up in the second hydrophone. Clearly sound was being emitted in front of the animal, but proof was needed that it was being used for echolocation. It was necessary, therefore, to blindfold a dolphin in order to determine whether it could use sound alone to get about in it tank. The researchers had many failures for its is very difficult to tie anything on or make anything stick to a dolphin. The solution to the problem was a pair of rubber suction cups, one placed over each of the dolphin's eyes. The test animal, a bottlenose dolphin named Kathy, swam off across the tank blindfolded as if nothing was wrong. With ease she picked her way through a maze of poles without ever touching one, and was able to locate small objects on the far side of her ten metre tank. Small pieces of fish dropped right next to a barrier in the maze would be scooped up without touching the obstacle. Another test with Kathy was a discrimination test. She was invited to distinguish between a horse capsule filled with water and a piece of fish about the same size. Kathy took the fish every time. Dolphins clearly can use a sense other than sight to navigate and locate food. As dolphins have all but lost their sense of smell, and extra-sensory perception has zero scientific credibility, sound was considered to be the likely candidate.

One of the problems, though, is understanding how dolphins make these sounds. They don't have vocal chords and are rarely seen to blow bubbles when vocalising. This is a controversial area of research. One school of thought has held that the sounds are produced in the larynx just as in other mammals, but echolocation clicks seem to emanate from the forehead rather than the throat. Indeed, if microphones are placed around the head of a 'clicking' dolphin the sounds can be triangulated to deep in the forehead, at the back of the nostrils. If probes are placed in the muscles of the larynx and nostrils it has been found that during sound production the larynx is quiescent while valves in the nostrils show muscle-activity. Ultrasound scans have given similar results. At Boston University, R. Stuart Mackay and H. M. Liaw projected narrow beams of low-intensity ultrasound at a frequency too high for the dolphins to hear and were able to identify the structures that moved during sound production. The apparatus was a modified foetal heart monitor of a type found in most maternity hospitals. The observers were able to see the nasal plug and the vestibular, nasofrontal and premaxillary air sacs vibrate with clicking or buzz sounds. Nasal diverticula on the right side vibrated all the time when clicks were produced while the left nasal diverticula vibrated only some of the time. The vestibular sac inflated as the clicking sound was made, probably as a resonator. The nasal plugs were thought to be the site of the original sound production as air moved upward, presumably from the lungs. When the blowhole is closed, air recycles to the vestibular sac. Clicks seem to originate in the right diverticula. Whistles are thought to be generated in much the same way as human whistles except that they are 'blown' internally.

Dolphins have a pair of nostrils inside the head which come together under a single blow-hole. The land ancestors of these animals probably possessed paired external nostrils, just as most mammals do today, but as dolphins evolved into divers they needed a way of storing air if they were to make and use sound underwater. They developed a covering over the nostrils, the blow-hole, with a complicated series of nasal sacs and valves below. Air is breathed in at the surface, and on diving, the blow-hole is closed. The air trapped in the dolphin's respiratory system can then be passed across the valves in the nasal sacs to produce sounds and then recycled through a complicated series of piping to be used over and over again as the dolphin echolocates underwater. The click sounds are thought to leave the dolphin's body through the forehead. At the front of the forehead is a large fatty body known as the melon which focuses the sound, much as an optical lens focuses light, about a metre in front of the animal's head. Donald Malins and Usha Varanasi, of Seattle University have found a concentration of unusual lipids, made of isovaleric acid, at the centre of the dolphin's forehead. The three-dimensional arrangement of these small molecules, rarely found in the lipids of other animals, suggests to the researchers that the area is, indeed, a 'sound lens'. When the sound returns, having bounced off a target, it does not enter an ear canal but is picked up by the thin bone of the lower jaw as a vibration, travels along fatty tissue in the hollow lower jaw, and thereby is transferred directly to the middle ear. Sound production, transmission, and reception is optimised for life underwater.

The rapidity with which the clicks are emitted means that humans cannot hear individual packets of sound, rather we hear trains of pulses much like a creaking old door hinge closing. One of the remarkable things about dolphin

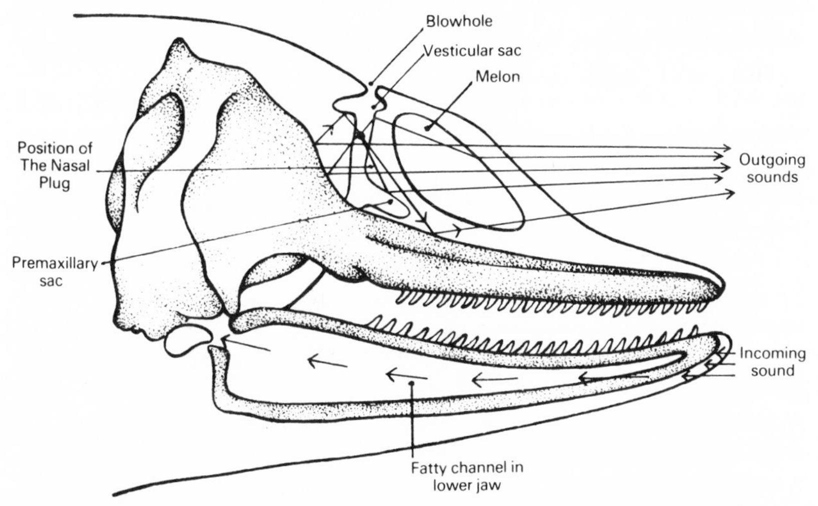

Fig. 13 Simplified section of dolphin head showing structures associated with sound production and detection. (After Norris)

echolocation is the rapidity with which the ear and brain process these clicks. A click-train may be made from up to 700 units of sound per second. A dolphin is capable of mentally separating these units, listening to the individual echoes, and decoding the information while interrogating a target. In the human ear the sounds would fuse together in our minds at 20–30 clicks per second. The sound is also known to enter solid structures. In experiments, dolphins have been able to tell the difference between a copper and aluminium plate painted the same colour. They also have the ability to distinguish a hollow aluminium tube from a solid one, although both tubes looked identical from the outside.

Perhaps one of the most interesting aspects of dolphin research centres on the sound beam itself. Modern dolphins have narrow beams of sound; the bottlenose dolphin, for example, has a beam width of about 9°, a thin pencil-beam of sound. The Indus river dolphin *Platanista indi*, a more primitive species, emits a 65° beam. Part of the evolution of these creatures might have been a narrowing of the beam width, giving a greater distance penetration for the same amount of energy. The concentration of energy in narrow-beam dolphins has become so intense that the prey being detected has become affected by it, leading to an entirely novel way of catching lunch.

Together with Bertel Mohl, of Aarhus University, Denmark, Ken Norris has been pursuing the idea that some of the toothed whales may have the capacity to debilitate their prey with sound. The sounds put out may be so intense that the prey may be killed, or at least immobilised to prevent escape. The idea is not new. Drs V. M. Bel'Kovitch and Yablokov, in the USSR, calculated that dolphin sounds should have enough energy to incapacitate prey. Further support came from the work of Dr A. A. Berzin with sperm whales. Berzin looked at a number of sperm whales caught by the whale fishing industry. Strangely, some of them came up with what looked like congenital deformities of the lower jaw. The lower jaw was curved so the animal was unable to close its jaws together to catch prey. In the stomachs, however, were plenty of squid and the whales looked otherwise perfectly healthy. Squid swim much faster than sperm whales, anyway, so it was doubly surprising that these enormous animals, the size of an omnibus, were able to take the ton or so of squid per day needed to survive. Berzin concluded that the jaws were not essential for feeding and that sound was being used to stun the squid.

In Hawaii, Whitlow W. Au and A. E. Murchison measured the intensities of sounds emitted by a bottlenose dolphin that was asked to carry out extreme-distance discrimination tests. A sphere, about the size of a tangerine, was placed about 113 metres from the dolphin. The animal was able to locate the sphere, but interestingly, when the echolocation sounds being produced were measured, they proved to be many orders of magnitude higher than any sounds recorded from a dolphin before. Indeed, they were close to the finite limit of sound, that is, the limit at which any more energy put into the water would simply turn to heat. With that kind of sound level Ken Norris asked the question – will those sounds kill prey? Tests with man-made sound beams showed that fish could be killed with a high-intensity sound beam. At the Marine Biological Laboratories at Plymouth, large squid could be killed quite rapidly by sound. In theory, then, dolphins could kill with sound, but the next question was whether they used this weapon. Norris considered that it would be odd for a creature to have evolved such a weapon if it didn't use it. Further tests were carried out with dolphins in captivity. Live fish were placed in a tank with three Hawaiian spinner dolphins, and sure enough after a short while the dolphins began to spray the fish with sound. The fish used were much larger than the dolphins' normal prey so it was not expected that the dolphins would kill them. Norris and his colleagues, though, watched for any signs of debilitation. The dolphins continued to direct echolocation clicks at the fish school for over an hour. A dolphin would make a run at the school attempting to put the fish right on the tip of its beak. The fish school would split, head for the tail of the dolphin and reform. Slowly, though, the fish school became depolarised. The researcher noticed after a while that the fish were not all facing in the same direction. Then individuals began to wander away from the school, seeming totally disorientated. Similar observations have been made in the wild. Striped dolphins *Stenella coeruleoalba*, for example, have been seen to circle an anchovy school, spray them with sound, and then cut through the school shovelling fish into their jaws at will. The anchovies do not attempt to escape. It is possible that the fish eventually suffer from a build-up of waste products in the blood due to their strenuous efforts to escape but an interesting observation made by a biologist on a fishing boat off Vancouver does not appear to support that. A large salmon was clearly visible in the water below the boat. As the researcher watched, the salmon suddenly stopped swimming and directly behind a group of killer whales approached. One of the whales scooped up the salmon and swam on. The salmon made no attempt to escape. Another piece of anecdotal evidence is that many scuba divers in dolphin circuses have felt a light touching sensation on the backs of the necks which they have attributed to the echolocation activity of their charges.

If the interpretation is correct, dolphins are clearly formidable killers, but one problem individuals in a large dolphin school might have is the danger of zapping another dolphin. In a school of actively feeding dolphins it would be very easy for one animal seriously to upset another, and dolphins are known to get angry with each other. To look for signs of 'echolocation manners', Ken Norris watched and listened carefully to three individuals in a tank. In more than a hundred crossings of one dolphin with its fellows, it never once echolocated them. As a dolphin crossed in front, the actively echolocating animal switched its gear off, turned its head away, and then switched back on again. With such a devastating weapon in the dolphin's head, it is important that 'echolocation manners' are a part of the social organisation of a dolphin school.

Large schools of dolphins may consist of hundreds or even thousands of animals which hunt in a 'line-abreast' formation. In this way a broad expanse of ocean can be echolocated in the search for food. Sound is thought to be used also in locating good food-gathering areas. Dolphins, it seems, can listen to the underwater background noise associated with the increased number of organisms at submarine escarpments and sea-mounts. There is also evidence that small schools of bottlenose dolphins sometimes detect and follow the sounds being made by pilot whale schools. The pilot whales, with their longer diving times, appear to be better at finding food and the dolphins take advantage of this superior food-finding ability. Having located a school of fish, a large school of dolphins will change its shape during the attack although they will always dive in synchrony. Sound signals must keep the school in step. Killer whales have been seen to work together in a similar way to capture prey. Observers of harpooned baleen whales off Australia have described pod members hanging onto lips, flukes, and lying on the blow-hole in attempts to immobilise the whales. On other occasions killers will encircle prey animals like seals or walruses, coordinating their moves by sound. When approached by killer whales, other cetaceans often become silent, flee quickly or 'spy-hop', that is they rise out of the water and search the surface.

4

CONSTRAINTS OF THE ENVIRONMENT

In the tropical rain forest, insects, amphibians, birds and mammals compete to be heard in the cathedral-like gallery of towering, broad-leaved trees. In order to achieve long distance communication within a jungle, information gained through hearing is probably more important than that gained by sight. Territoriality is one obvious context. In a large territory in a forest, sound will allow the territory holder to proclaim his residency without having to visit all the sectors he controls. If many creatures are calling, all at the same time, the sender must have a sufficiently different sound from the rest in order to get through to the receiver. There is also non-biological competition from gurgly streams, wind rustling the leaves in trees, waterfalls and maybe waves crashing on a nearby shore. Then there are atmospheric changes in temperature, pressure and density. To overcome these constraints of the environment birds, for instance, have developed their characteristic songs and calls. An almost pure tone musical song will stand out from wide-band background noise. Give the notes a rhythm, a temporal pattern, and they differ sufficiently from non-animate sounds to be picked out loud and clear. Imagine an animal which is only capable of producing a simple whistle as its sole means of sound communication. If it encodes its signal by making the sounds alternately loud and soft it would find it hard to communicate over long distances, because, as the wind blows and the environment changes, the whistle pattern would be degraded. Sometimes, when the signal was meant to be quiet, it might come through loud, while at other times a loud signal might appear unpredictably soft. A listener would be totally confused as the pattern received would not correspond to the one that was sent. If, on the other hand, the whistle was sung long and loud, followed by silence and then again short and loud, this pattern would always get through. An elaborate, rhythmic, musical song is the means some birds have acquired in order to talk over long distances to other individuals of the same species. What they are saying we leave for later chapters, but for now let us consider environmental constraints and acoustics, in particular those in a forest or wood.

Vegetation in a forest – the leaves, branches and trunks of trees, fallen logs, blades of grass, flowers, bushes and leaf litter – can dramatically affect the quality of an animal's call, making it difficult to transmit the signal to any useful distance and sufficiently unchanged to allow a receiver to get the original and intended message.

Perhaps the first thing to be considered is one of nature's basic physical laws. In a hypothetical, homogenous and frictionless medium, sound will travel out from the sound source in the shape of an expanding sphere; that is, it will spread out equally in all directions. The energy is more thinly spread as the sphere enlarges and the inverse square law tells us that for each doubling of distance there will be a drop of 6 dB in sound intensity. The sound signal is said to attenuate, which simply means that the loudness is gradually reduced as it moves away from the source. In air, a bird's song not only attenuates, but the frequencies contained in the song attenuate at different rates. High frequencies lose their energy faster than low frequencies. A signal containing both high and low frequencies, therefore, after travelling over a long distance, will reach the receiver in a slightly different form from when it left the sender.

Natural environments differ a great deal from the theoretical model. There are things in the way that alter the course of events. In addition to the air itself, there is vegetation, temperature change, wind movement, eddies and currents, density gradients and interference from other callers. A forest is a particularly complex acoustic environment compared, say, to a flat grassy plain and animals adapt their songs and calls to suit. Dr Eugene Morton, a research zoologist at the National Zoological Park of the Smithsonian Institute, Washington, DC, undertook a comparative study of the acoustics of forest, forest-edge, and grassland habitats of Panama.

He first looked at sound transmission, using pure tones to determine which frequencies travelled best in a particular habitat. In the forest a narrow frequency range or 'sound window' of between 1,585 and 2,500 Hz attenuated least, and therefore propagated best, but only if the transmitter and receiver were placed about five feet above the forest floor. When put directly on the ground, sound was rapidly absorbed. If sender and receiver were more than 100 ft apart, reflections from trees and leaves altered the sound appreciably, although reflection from the canopy did mean that some sort of signal got through. On the forest edge the absence of the canopy meant that sounds were not deflected back to the ground and lost. In the grassland habitat there was no appreciable sound window and reflections played little role in distorting or attenuating the sound transmission.

Morton then taped as many of the bird species from each habitat as he could find, augmenting those recordings with examples of the species he couldn't find which were stored at the Laboratory of Ornithology of Cornell University, Ithaca, New York. Each of the songs was analysed for several components. An important factor to be identified was the dominant frequency, the one where most of the energy in the call was concentrated. Another component was the type of sound, whether a tonal whistle or a rapidly modulated buzz.

In the forest, the birds living at lower levels sang with an average dominant frequency of 2,200 Hz, which conveniently fitted the sound propagation results. The normal song type was a tonal whistle. In the open grassland, where there was no sound window, the birds sang with high frequency, modulated or buzzy sounds. Along the forest edge the birds were divided equally between the modulated and tonal song types. It occurred to Morton that forest birds relied mainly on changes in frequency in order to call long distances, while those on the plains were using the temporal pattern to encode signals.

Open grasslands, although acoustically less complex because of the absence of thick foliage, are also difficult habitats for sound transmission. Wind, temperature and turbulence disturb the signal. There is also a curious 'sound shadow' near to the ground due to temperature changes and air movements at that level, and to absorption by the earth. A bird must get itself ten to 20 ft above the surface to enhance its broadcast area. Eugene Morton noticed that a quarter of the grassland birds in Panama sang while in flight. Other workers have recorded that flight songs are more prevalent in birds living on temperate grasslands and tundra. Morton also noted the times of day at which grassland species called and the type of call they used. The rufous-tailed hummingbird *Amazilia tzacatl*, for instance, is an unusual grassland bird in that it has a tonal, high frequency call. It only sings between six and seven o'clock in the morning, just before the air has had time to warm up. The fork-tailed emerald *Chlorostilbon canivetii*, on the other hand, buzzed contentedly for the best part of the day. Interestingly, Morton identified one group of wedge-tailed ground finches *Emberizoides herbicola* that sang tonal songs just before the sun came up, while another group of the same species nearby sang throughout the day but with a buzzy song. On the other side of the Atlantic, on the African continent, the French ornithologist, C. Chappuis, came to similar conclusions although using a different method from Morton. Chappuis simply relied on his own perception of how loud sounds appeared at various frequencies.

Further south, on the island of Madagascar, Lee McGeorge, working from Duke University, North Carolina, looked at the way vegetation and background chatter in the Berentz gallery rain-forest influenced the calls of the animal community living there. She also undertook sound propagation tests, at a height of 1.7 metres, in order to identify sound windows. In the forest, two windows were identified, one between 1,000 and 2,000 Hz, and another between 4,000 and 5,000 Hz. The lower sound window contained the average emphasised frequencies of many of the long-distance calls of Malagasy animals, including the 'mew' call of the ring-tailed lemur *Lemur catta* and the 'yodel' of the giant ground coua *Coua gigas*, but excluding the song birds. The small nocturnal weasel lemur *Lepilemur mustelimus* emits its loud territorial defence calls in the higher sound window. It is thought that higher frequencies travel better at night since the atmospheric conditions are more favourable then. 'At night', according to Carl Eyring, from Brigham Young University, Utah, 'the low frequencies decrease as the light breezes cease and the high frequencies increase as the insects begin their nocturnal chorus'. In McGeorge's Malagasy forest, other night-time creatures like the mouse-sized lesser mouse lemur *Microcebus murimus* and the fat-tailed dwarf lemur *Cheirogaleus medius* call at frequencies above 4,000 Hz. During the day, however, the lower the frequency the better the call is transmitted. Eyring, working in a Panamanian rain-forest, noted that attenuation of the higher frequencies increased with humidity.

Songbirds call all day, so McGeorge was curious to find that the Malagasy passerines sang with dominant sound frequencies ranging anywhere from 500 to 7,000 Hz. She suggested that the wide range of calls might be related to the number of animals trying to use the communication channel. If they all tried to call at the optimum frequencies, background noise in the sound window would be unacceptable. Another way of overcoming this kind of congestion is to space your call differently from your neighbours, or to sing at different times. McGeorge found that not only is there diurnal spacing of callers on the same frequencies but also a similar seasonal spacing.

There is also evidence from monkey studies for animals taking advantage of sound windows. Peter Waser, for his thesis at Rockefeller University, studied African mangabeys. The sun in the early morning hits the forest canopy which rapidly heats up. As a result there is a steep density gradient high up at the roof of the forest. If a monkey produces a loud sound somewhere below, half-way up a tree, then there is a curious channelling of the sound. Instead of the sound passing through the canopy and dispersing, it gets refracted back down without loss of energy and there is transmission over much greater distances than would be possible at other times of the day.

Sound windows have also been suggested for northern temperate deciduous and coniferous woodland. T. F. W. Embleton, in the USA, showed that cedar, pine and spruce leaves seem to enhance the transmission qualities of sounds between 1,000 and 3,000 Hz. In deciduous woods the window is between 1,000 and 2,000 Hz. Many factors have been identified as affecting sounds as they travel through a temperate forest. There are varying degrees of absorption depending on the nature of the soil surface. Embleton considered tree trunks and stems as cylindrical objects in a forest and measured their surface impedance. G. Bech, at Berlin University, glued overlapping leaves onto frames, and measured the amount of sound the leaves absorbed. At Nijmegen University in The Netherlands, Maurice Martens went even further in order to investigate the influence of foliage on sound transmission through vegetation. His experiment was carried out with model forests placed in an anechoic chamber. First he grew birch trees, hazel trees and privets in earthenware pots. When they were of sufficient size he placed 46 birch trees in the chamber (experiment d) and played white noise across this artificial forest. With the aid of microphones placed in front and behind the miniature trees he could measure the amount of sound going in and the amount passing through and coming out at the other side.

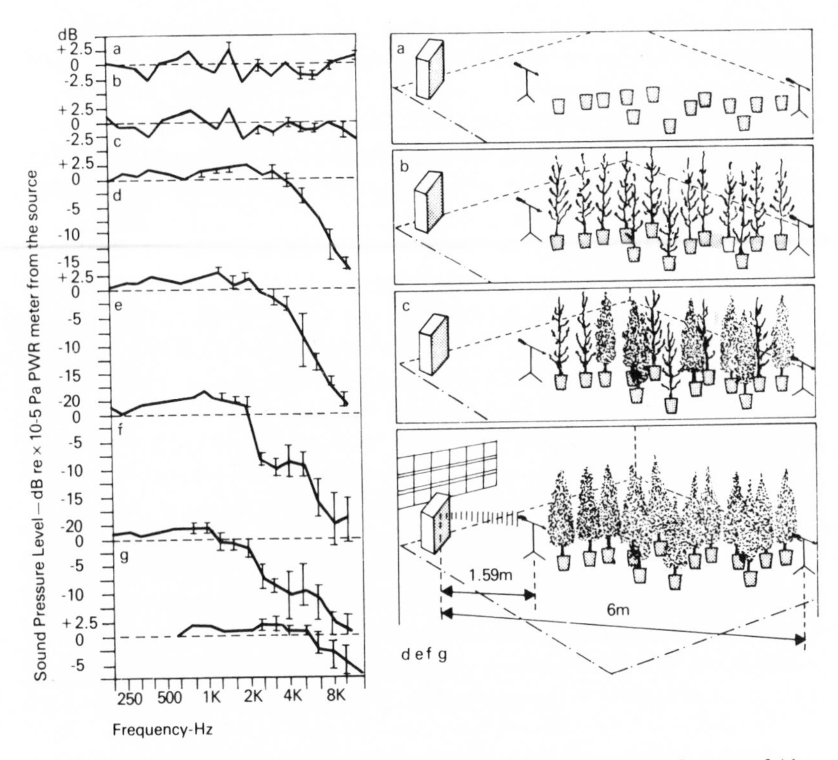

Fig. 14 Schematic view of experimental arrangement testing the influence of 46 flowerpots and trees on the noise field in an anechoic chamber: A. Birch trees sawn down. B. Birch trees defoliated. C. Birch trees half of which are defoliated. D. Fully foliated birch trees. E. Hazel. F. Tropical plants. G. Privet.

He repeated the experiment with half the trees defoliated (experiment c) then with all the leaves defoliated (experiment b), and finally with the plants sawn down and just their stumps remaining in the pot (experiment a). Similarly, 25 hazel trees, 12 privets, and a selection of tropical plants were tested separately. Martens was able to show that, between 200 and 12,500 Hz, foliage influences the sound field, and that in general in the mid-frequencies the vegetation acts as an amplifier, boosting 200–3,200 Hz in birch, 200–2,000 Hz in hazel, 200–1,000 Hz with tropical plants and 640–5,000 Hz in privet. Martens concluded that the acoustic properties of the vegetation could be 'important as an evolutionary environmental factor in the development of communication and vocalisation of animals living in vegetative habitats'. In temperate deciduous forests the picture is further complicated by seasonal variations. Why the amplification phenomena occur at all is still unexplained, but one positive thing did come out of the study – Martens suggested that the filtering property of plant communities might be useful in noise abatement.

A team of researchers at Pennsylvania State University's Noise Control Laboratory, investigating the effectiveness of trees and woods as traffic mufflers, made use of forest acoustic studies. The leader of the team, Dr Gerhard Reethof, proposed that sounds are not absorbed by the trees themselves, but by the leaf litter and humus layers on the forest floor. Thus wooded areas would be better acoustic mufflers than open parkland and mini-urban forests would make living near busy streets and motorways more comfortable.

Back in the forest, songs are not only affected by absorption and filtering but also by reverberation or echoes and fluctuations in amplitude or loudness. Reverberation is caused by sounds reflecting off the surfaces of leaves, twigs, and tree trunks. Fluctuations in loudness may be a result of the sound passing through air of different densities. There might be eddies and turbulence in the atmosphere; also temperature differences and, of course, wind. The effects of these environmental constraints can be quite serious for a bird song of rapidly repeated units. Reverberation will make the units blur into each other. The pattern of song will be lost and any bird listening for the message would be hopelessly confused. The problem of reverberation increases with a rise in frequency because smaller wavelengths are reflected from more surfaces, particularly in a forest. One of the major problems for birds is that they produce high frequency songs and calls. Frequencies in the region of 4,000–6,000 Hz are severely affected by reverberation over a distance of 100–200 metres in a forest. Frequencies below 1,000 Hz are much less affected by reverberation. Someone who has been working on the effects of reverberation and amplitude fluctuations on bird songs is Haven Wiley, from the University of North Carolina, and from the results of his experiments he is beginning to question the evidence that birds are using sound windows.

If sound windows were being used by birds for optimal song transmission, then they would be using frequencies as low as possible to avoid reverberation and other distortion factors. It is now not clear, suggests Wiley, why birds use the frequencies they *do* use. If birds want to be heard as far as possible for the least effort then low frequencies have all the advantages. Clearly small birds have a simple physical problem of not being able to generate low frequency sound when they have only a very small sound generating mechanism, the syrinx, with which to work. Given that birds are handicapped by their size, Haven Wiley undertook to identify strategies that birds might use to make correct recognition of their acoustic signals easier for a receiver. There are several possibilities. An increase in loudness will increase the contrast between the intended signal and the background chatter. A repetition of the signal, although a form of redundancy, will probably mean that at least one signal will get through. A third strategy might be to reduce the repertoire of signals. It is easier for a listener to expect and to spot two signals, rather than three. Lastly, the use of an alerting component might get receivers in a state ready to pick up a subsequent message, by knowing the exact moment it is expected to arrive.

The longer the distance between sender and receiver, the more degradation and attenuation the signal will encounter before the receiver is confronted with the problem of detecting it. Reverberation will blur signals which carry information in the spacing of notes, such as rapid trills, whereas amplitude fluctuations caused by, say, turbulence, will mask slow amplitude patterns, such as long melodic songs, so the signal will come and go like a bad radio signal. One prediction that could be made, therefore, is that acoustic signals for long range communication ought to rely on frequency patterns rather than amplitude patterns to encode such messages as species identity or individual identity and this is precisely what bird song does. Bird songs are remarkable for their complex patterns of frequency modulation and the birds themselves are insensitive to many sorts of alteration in the amplitude patterns of their songs.

During evolution most birds have ignored amplitude patterns. But this is not true of other animals. Forest primates, for example, which have long range vocalisations to call back and forth between troops which may be hundreds of metres apart, have songs with patterns of repeated pulses of sound. Sheer redundancy or repetition of the pattern eventually gets the signal through. There are, however, some mammal signals that *do* use frequency patterns; the whistling of ground squirrels and bull elks, for example, are both tonal, frequency modulated calls.

A second prediction is related to the differences between major categories of habitats. In forests, with their abundance of reflecting surfaces, reverberations will be much more severe than in open environments, whereas amplitude fluctuations will be greater in more open habitats. Winds are generally stronger in the open and temperature gradients are more marked. The heating of an open plain by the sun, for instance, will result in rising columns of warm air. Sounds travelling across such a turbulent plain would be distorted. Birds in forests, therefore, should avoid rapid repetitions of any one frequency component, whereas those in the open ought not to use slow melodic acoustic patterns. Haven Wiley investigated the birds in North Carolina, and it turned out that most birds comply with the prediction. In North America the various species of wood warblers, like the hooded warbler, breed in dense woodland and lack rapid trills in their songs. In the open habitats, various pipits *Anthus spp.* and the horned (shore) lark *Erenophila alpestris* have rapid, almost jangling notes in their songs. In Europe, the nightingale *Luscinia megarhynchos* possesses the characteristic song of deep forest birds. Wiley's observations had revealed similar acoustic patterns in the bird song of his North American birds to those of Eugene Morton's birds in Panama. The forest had its 'tonal singers', while the grassland had the 'buzzers'.

In Europe, John Krebs of Oxford University looked at a species of bird that could be found in both woodland and open grassland. If Wiley's prediction was right, great tits *Parus major* in forests should have a different pattern of song from those in open parks. The great tit has a wide distribution, in diverse habitats, from Japan in the east to Ireland in the west, and from Siberia in the north to Malaysia in the south. Working with vegetation maps, John Krebs and his colleagues were able to identify geographical regions of large homogeneous blocks of either dense forest or open woodland and sample the songs of the birds singing there. Returning to Britain they analysed the songs for pattern and emphasised frequencies. Great tits living in forests, whether in Britain, Poland, Norway or Morocco, had simple songs narrow in frequency range, with a relatively low maximum frequency and a slow repetition rate of notes. In contrast, those in arid woodlands or parklands had a wide frequency range, high maximum frequencies, and a higher repetition rate. Krebs' observations in Europe with great tits had supported the observations of both Morton in Panama and Wiley in North Carolina. Birds, it seems, adapt their songs and calls to suit the habitat in which they live.

But, if their vocalisations are inevitably going to be degraded, could it be that birds actually evolve song structures that permit some degradation for the express purpose of allowing receivers an easy mechanism for judging the distance at which a singer is located? In evolutionary terms it might prove to be advantageous for birds in certain circumstances to be able to judge distances by the quality of song. If a receiver could not easily locate and judge the distance of a singer, when it came to marking out its own territory, for instance, the only way it could determine the singer's territorial boundary would be by a short range encounter which might be more traumatic for both parties than long range vocal avoidance.

To test this hypothesis, one of Wiley's colleagues, Douglas Richards, carried out some clever experiments with Carolina wrens *Thryothorus ludovicianus.* The bird was chosen because of its characteristic response to experimental playbacks of songs. If a song is played back inside the territory of a Carolina wren, the resident bird immediately stops its own singing, approaches the loudspeaker uttering aggressive calls, and may even physically attack. As long as it judges that the playback is within its territory it will not sing. If, on the other hand, the song is played back outside its territory the resident responds by singing to the playback. Could a Carolina wren, therefore, tell whether a singer is inside or outside the territory solely by judging the distance on the basis of some predictable degradation in the song pattern as a result of propagation through a forest? Carolina wrens in the wild were recorded at two distances, one at ten metres and the other at 50 metres. The recordings were filtered to exclude frequencies above 3,000 Hz to minimise frequency dependent attenuation as a cue for judging distance. The recordings differed mainly in the amount of reverberation. Those recorded close to a wren were relatively clean, while those 50 metres away suffered a great deal. The experimenters were careful that the playbacks were at the same amplitude. The recordings were then played back to Carolina wrens on their territories at a standard distance of 25 metres. This was close enough to be inside the resident's territory, and would normally result in the termination of singing and an approach. The resident wrens, however, responded quite differently to the two recordings. They stopped singing and approached the clean recordings, but the degraded recordings, broadcast at the same volume and at the same distance, evoked counter-singing. Wiley and Richards, therefore, were able to conclude that Carolina wrens, at least, could use the degradation that naturally occurs in the transmission of their songs through a forest to judge the distance to a singing bird.

Another factor which interested Wiley and Richards was the possibility of birds using an alerting component in their songs. A bird has to live, to feed, to watch out for predators, and to attend to its mate or offspring. It does not have time to hang about waiting for some other bird to call to it. It must be alert to potentially useful messages in the general background noise of the forest, yet not be preoccupied with communication. A receiver could operate with a low level of vigilance if any interesting incoming message was provided with an initial alerting signal. This would have to be an easily detectable part of the song or call, which suffered low degradation, and had high contrast with background noise. Once the alerting component is detected, the receiver could switch its attention to a brief 'time window' that it knew would contain a message component which might be more complicated and contain much more information, such as species or individual identity. Many bird songs appear to be designed in this way. They start with a simple tonal element, often a simple whistle or series of whistles repeated at a slow rate, then, the acoustic structure becomes more complex with rapid trills and so on.

Again, Douglas Richards designed an intriguing experiment to test for alerting components. The subject on this occasion was the North American rufous-sided towhee *Pipilo erythrophthalmus* which begins its song with one or two simple whistles and then follows with a complex trill. Towhee song played back within a resident's territory will evoke counter-singing. Richards played back different versions of towhee song, including the suspected alerting component only, the message component only, and the complete song. Some of the recordings were degraded to simulate noisy and reverberant forest conditions. If the degraded opening whistle or the degraded trill was played back to the resident bird there would be little response. A clean trill, without reverberation, evoked a strong response from a territorial towhee. The key experiment came with playbacks of clean and degraded full song. A clean version of the entire song, not unexpectedly, received a strong response, but what was exciting was that so, too, did a degraded version. Wiley and Richards were able to conclude that an alerting component in towhee song *does* allow another towhee to recognise the signal as coming from a rival towhee, and permits the receiver to get the message in the face of high levels of degradation and noise. So, in that one species at least, there is evidence that the acoustic pattern of a bird's song is adapted very nicely to improving signal detection.

In general, animals clearly adapt their songs to suit their environment in order to ensure optimum propagation of their signal. When a species gradually, over tens or maybe thousands of years, migrates from, say, field to forest, the song becomes inappropriate and a new song evolves. What then happens if habitats are lost more rapidly? Can animals adapt quickly enough to overcome the constraints of the environment imposed upon them today? With tropical forests, for example, disappearing at the rate of 20 hectares a minute, even a simple task like talking to another of the same species suddenly becomes a problem, and may even become impossible.

5

BIRD CALLS AND SONGS

In the harsh and stormy south Atlantic, close to the northerly limits of Antarctic pack-ice, lies a refuge for sea birds. In summer, Bird Island, half a mile off the coast of South Georgia, becomes home for breeding colonies of thousands upon thousands of sea birds. On this tiny island (three miles long and half a mile wide), with its mountains, glaciers, fjords, rocky beaches and sea cliffs, there can be found a multitude of species – wandering albatrosses, black-browed albatrosses, light-mantled sooty albatrosses, grey-head albatrosses, macaroni penguins, king penguins, Dominican gulls, Antarctic terns, prions, southern and brown skuas, sheath bills, blue-eyed cormorants, giant petrels, South Georgia brown pintails, and Antarctic pipits.

It is thought to be the largest concentration of biomass in the world. The sea birds pack onto this tiny speck in the ocean, nesting on every inch of space. The movement of individuals is confusing, and the noise they make is deafening. For a pair of birds to find each other in the general mêlée, somehow they must recognise each other at a distance. Many birds have visual powers so acute that one can spot its mate at a distance of 200 metres. Even in the difficult conditions of a dense breeding colony, however, sound recognition appears to be important.

Bird Calls

In an attempt to identify the relative importance of sound and vision in sea bird recognition, Professor W. H. Thorpe of Cambridge University, a pioneer in the study of animal sounds, visited the Bass Rock in the Firth of Forth. There he was confronted with an enormous colony of noisy gannets *Sula bassana.* Even in the gannet's simple squawky call, Thorpe found that there was enough information for the individual to be recognised by others in the colony. While recording gannets on their craggy cliff-top nests he noticed that, when one of a pair returned with its catch, it would fly to the bottom of the cliff, hang in the up-draft and move up the cliff face as if going up in a lift, calling all the way. The gannet on the nest paid no attention to the rising current of birds until it heard the returning mate. It became excited, and began calling back. Thorpe concluded that each bird could recognise the calls of its own mate as distinct from those of others. He found that only the first part of the call, the first tenth of a second, was needed. A similar study was carried out with Sandwich terns *Sterna sandvicensis*; this time the way chicks spotted incoming parents was studied. The parent terns locate the nest site with the aid of their visual memory. They give a 'fish call' when approaching the area, but only their own offspring respond. The young quite obviously recognised, Thorpe discovered, three distinct sections in the call of the parents.

One of the first scientists to show that birds were able to recognise each other by sound alone was B. Tschanz. In 1968, he published a paper which showed that individual guillemots *Uria aalge* produced physically distinct calls to which their young or mates reacted significantly more than to the calls of others. Tschanz carried out playback experiments. Young birds in the nest were exposed to parents' calls and control calls, alternately and simultaneously, using two speakers. Young guillemots paid no attention to the control calls and responded only to the calls of their parents. Indeed, they would peck at, and beg from the loudspeaker playing parental calls. Tschanz also went back a stage in the birds' development and exposed eggs in incubators to particular adult calls. After hatching they were played the adult calls and other control calls. They responded to the calls to which they had been exposed while in the egg. An auditory signal can be learned even by the embryo before hatching.

Individual recognition is only one function of bird calls. Social calls may keep a flock together. Alarm calls warn others of danger. Feeding calls are exchanged between mates. There may be roosting calls, nest site calls, and so on. Calls can be associated with particular events in a bird's life. Aggressive

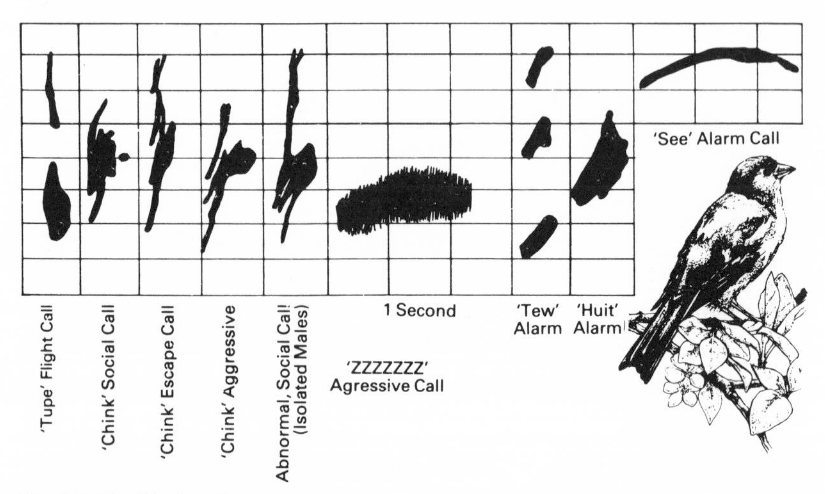

Fig. 15 Chaffinch call sonagrams.

situations might give rise to a threat call. Youngsters give begging calls for food. Thorpe worked out the vocabulary of calls for a variety of species of birds, and found that the maximum number of calls a bird might have is around 15. The passerine birds came top of the list.

Bird calls tend to be simple, almost monosyllabic, and relatively easy to understand. A blackbird gives a 'pinking' call in response to the presence of the neighbourhood cat. The meaning is clear.

Nature of Call	*Call*	*Remarks*
1 Flight call	'Tsup' or 'Tupe'	A short penetrating call. Low pitch with single higher harmonic. Associated with flight or preparation for flight.
2 Social call	'Chink' or 'Spink'	Clear ringing call in two parts. Helps separated birds to meet again.
3 Escape call	'Cheenk'	More shrill than 2. Used as escape call and during courtship by newly-paired males
4 Aggressive call	'Zzzzzz' or 'Zh-zh-zh'	Low buzz uttered during attack and fighting in a few captive males.
5 Alarm call	'Tew'	Most frequent of three alarms. Common in young birds and more rarely adults of both sexes.
6 Alarm call	'Seee'	Extreme alarm in breeding male. Pure tone rising and falling and difficult for predator to locate precisely.
7 Alarm call	'Huit' or 'Whit'	Commonly at rate of 30 per minute. Male chaffinches in spring. Moderate danger.
8 Injury call	'Tseee'	Squeak given by birds hurt fighting.
9 Courtship call	'Kseep', 'Tsit', 'Chwit', 'Tzit'	Short and high-pitched in bursts of three simultaneous descending notes. Male gives it courting female at pair-formation in early part of season.
10 Courtship call	'Tchirp' or 'Chirri'	Coarse chirp. Replaces 9 during courtship later in season.
11 Courtship call	'Seep'	The only extra call of female during breeding season. Short and high-pitched but composed of only two notes.
12 Begging call	'Cheep'	Soft note of nestlings.
13 Begging call	'Chirrup'	Loud and penetrating call of fledgelings.
14 Intermediates	'Huit'/'Seee' 'Huit'/'Chink'	Occasional intermediates between alarm and social calls of mature birds.
15 Sub-song	'Chrrp' and variants or low-pitched rattles grouped together in various ways	Chirps and warbles of birds in first summer.
16 Song	'Tchip-tchip-tchip-Cherry-erry-erry-Tchip-Tcheweeoo	Trills and a terminal flourish lasting some two to three seconds. January or February to June, and September and October.

Fig. 16 The call vocabulary of the chaffinch.

Most calls are thought to be inherited rather than learned, and to be passed on unaltered from generation to generation. Recent research, however, has revealed that things are not always as simple as this. Paul Mundinger, at Queen's College, New York, studied the North American goldfinch *Carduelis tristis.* One of its calls, a flight call, is used by members of a pair to maintain contact during the breeding season. The female goldfinch has the habit of incubating the eggs for long periods of time. The male feeds the female. As soon as the male returns to the nest area, the female begins to beg. Mundinger, hidden near the goldfinch nest, noticed that the female began the begging behaviour long before she could see the male. He concluded the female had heard the flight call that the male produces during its 'bounding' flight. Mundinger recorded the flight call and analysis revealed it to be more complicated than many other bird calls. In addition, individuals had subtly different variations. The next spring, Mundinger was able to show that, at the beginning of the breeding season, a pair of goldfinches will modify the structures of their flight calls to a common pattern. One bird, the male, in fact, imitates one of the calls of the other. They maintain this pattern throughout the breeding season, using the calls as a kind of naming system.

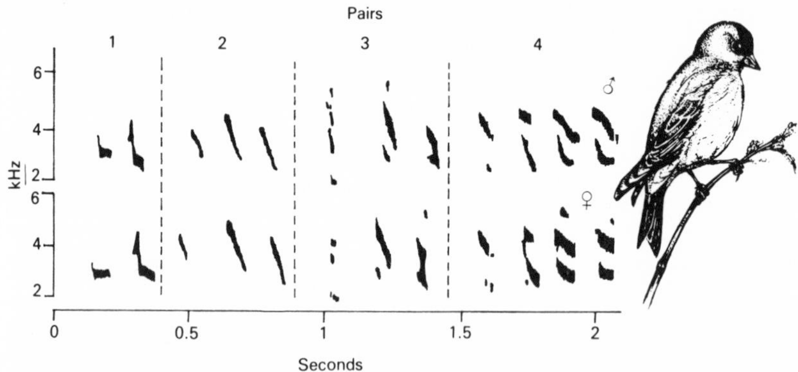

Fig. 17 American goldfinch pairs modify their calls to a common pattern.

He was also able to demonstrate that this capacity to learn new flight calls is maintained throughout life. Mundinger had male goldfinches as old as nine years still learning new flight calls when each new breeding season came around. The advantage to the male is that it can establish a bond with a new female if a mate dies or moves away. Further research has indicated that, in the family of cardueline finches, many calls in the repertoire are learned. Finches reared in acoustic isolation produce calls which seem normal to the human ear. The modifications through learning are quite minor and involve detail in call structure. Nonetheless, the modifications are sufficient for individual finches to discriminate one another's calls. In Norway, Paul Mundinger and Peter Marler discovered, similarly, that mated pairs of twites *Acanthis flavirostris* have calls in common, which are different from the calls of other pairs, and which appear to be learned. They use the calls to make contact when in flocks.

At the University of Sussex, Peter Slater noticed a similar learned component in certain chaffinch *Fringilla coelebs* calls. In some of his hand-reared chaffinches the 'chink' call is quite different from the same call given by wild birds. Having not heard the 'chink' call of others, the hand-reared bird is at a disadvantage and consequently produces ill-rehearsed and variable sounds. Luis Baptista found dialects in chaffinch rain calls; birds in separate areas gave different calls. Peter Slater had a laboratory chaffinch duetting with a sparrow outside the building. The chaffinch would give the sparrow 'cheep' when the sparrow did so. Clearly, birds are able to copy calls, even the calls of other species, and learning through copying is obviously necessary for the proper development of calls within a repertoire.

Calls between different species can sometimes convey information that would benefit individuals of other species. Alarm calls are obviously useful to all listeners. But the classic example of a bird using a particular call to attract, not a bird, but other animals, is the African honey-guide *Indicator indicator.* With a special call, the honey-guide leads animals such as honey badgers (ratels), or even local tribesmen, to the nests of wild bees. The honey-guide is not able to break open the honeybee nest and so it enlists the help of a larger animal. The badger, interested in obtaining the sweet honey inside, cracks open the nest, allowing the bird, which is more interested in the wax, to feed on broken pieces of the honeycomb. The ratel and the honey-guide were thought to be the original members of this symbiotic relationship. Man took advantage later. There is some evidence that in certain more urbanised areas, honey-guides have given up trying to entice men to help them, and concentrate solely on the badger population for their honey supplies. Honey, apparently, is more easily obtained at the local supermarket.

Bird Calls and Echolocation

Some birds, like bats and dolphins, indulge in autocommunication, communication with self. They bounce sounds off obstacles or targets in the environment in order to orientate and navigate.

The oil bird *Steatornis caripensis* of South and Central America and the West Indies is a kind of nightjar that roosts in caves. Its nest site is located on the cave wall. Each evening birds fly in and out of the cave in search of food. Flying in complete darkness, they produce a series of clicks. Echoes reflected from the sides of the cave walls enable them to find their way in the dark. Each click consists of a burst of pulses, the separation of which we cannot appreciate until the call is analysed spectrographically. The clicks are not ultrasonic and can be heard by the human ear. The noise created by a disturbed colony of oil birds in a cave is deafening. In experiments in darkened flight cages, oil birds have been shown to use their clicks, as do bats, to avoid obstacles in their flight path; if their ears are plugged they crash into things. When executing a complicated manoeuvre, such as landing on the nest site ledge, the bird's click rate increases, in the same fashion as the calls of the vespertilionid bats; the more clicks, the more information gained from the returning echoes.

Another group of echolocating birds is the cave swiftlets of South East Asia *Aerodramus spp.* The nests of these birds are used to make bird's nest soup, and are found in very dark caves. Swiftlets are diurnal, hunting for insects by day and returning to the caves at night. Their echolocation capability apparently allows them to stay out long after sunset, and to set out before sunrise, giving the colony a chance to exploit a larger area of food reserves than would be possible if they were using vision alone. Some species of swiftlets emit impulse bursts, like the oil birds, while others give pairs of clicks, similar to those of *Rusettus* fruit bats except that they are in the human hearing range. One species, *A. vanikorensis*, was found by Donald Griffin and Rod Suthers of Rockefeller University, to produce clicks between 4,500 Hz and 7,500 Hz. With these relatively low frequencies, the swiftlets can expect to receive a less detailed picture of their environment than can bats. Nevertheless, in tests *A. vanikorensis* was able to avoid 6 mm diameter wires in a darkened experimental room.

It is not known whether echolocation clicks also serve a normal, intra-specific, bird-to-bird communication function.

'Boomers'

Some birds produce very low frequency calls that can travel for long distances. The booming 500 Hz note of the bittern *Botaurus stellaris*, for example, can be heard several kilometres from the reed beds. In one case it was reported to have been heard five kilometres away.

The male American sage grouse *Centrocercus urophasianus* has an inflatable throat pouch with which it amplifies its booming leking call. The lek is an arena in which a group of male animals compete with each other for mating stations. Females visit the lek when ready to mate and are attracted most to the males occupying the central sites.

Another bird which researchers felt ought to be a 'boomer' but which, at first hearing, didn't appear to be is the capercaillie *Tetrao urogallus.* The cock bird visits the lek, fans out its tail, raises its head into the air and bursts forth, not with a deep boom, but with a rather insignificant series of clicks and pops. To Robert Moss, at the Institute of Terrestrial Ecology, Banchory, this seemed rather odd. The cock obviously puts a great deal of energy into the call and can be seen shaking all over. It inflates the enlarged oesophagus like a balloon, much as other grouse species do. With this apparatus the sage grouse, for example, can be heard over a kilometre away, but to human ears the capercaillie clicking, popping and squeaking call can barely be heard at 200 metres. Moss reasoned that, as the voice of the cock grouse is much lower than the hen's, so too should be the cock capercaillie's. The hen capercaillie has a deep voice, the cock's should be deeper, perhaps infrasonic. Moss went on to test this. Together with Ivor Lockie, of the Robert Gordon's Institute of Technology, Aberdeen, he recorded capercaillie calls and speeded the tapes up. They revealed that much of the sound is at frequencies below 40 Hz, which is inaudible to the human ear.

The second problem was to explain how the cock capercaillie achieves this kind of sound production. Physicists would explain that production of such a low frequency sound, on the organ pipe principle, would need an oesophagus of some great length, much longer than the size of the bird. An anatomical investigation showed that the capercaillie overcomes this problem in a rather clever way. It has a Helmholtz resonator in its throat. This is a closed tube, and when air is blown across the mouth, in the same way that a jug-band blower plays a jug, it produces a deep booming note, which, in the case of the cock capercaillie, is so deep we cannot hear it.

The rare New Zealand kakapo *Strigops habroptilus*, a species of nocturnal parrot which lives mostly on the ground, builds its own version of the Hollywood Bowl to project its very low frequency calls. Male kakapos gather in leks in the breeding season and boom, sometimes a thousand times an hour, for six or seven hours a night. The sound is amplified by air sacs that make them look almost spherical when calling, but this still doesn't account for the amplitude of the signal. It was noticed that the birds call from excavated hollows, once thought to be dust baths, but research has shown that these 'bowls' are the right shape, like an outside auditorium or amphitheatre, to reflect the sound far and wide.

'Drummers'

Some birds are not only vocalists but also instrumentalists. Coots *Fulica atra* stamp their feet in the water to frighten away rivals, peacocks *Paco cristatus* accompany their flamboyant visual display with stiff feather rattling, mute swans *Cygnus olor* have whistling wing beats that are thought to keep flying birds in touch with each other, and the woodpigeon *Columba palumbus* claps its wings together. But probably the most obvious instrumentalists are the woodpeckers, which drum on tree trunks with their strong beaks.

Dieter Wallschläger, of Humbolt University, East Germany, has been studying the woodpeckers' drumming behaviour and has found that, although birds drum on different trees and produce different quality notes, it is the temporal pattern which is important. Some birds drum fast, others slow, and yet others speed up during their performance. The great spotted woodpecker *Dendrocopos major* drums accelerando, whereas the lesser spotted woodpecker *D. minor* drums with constant time intervals, although these vary from winter, when it drums slowly at ten to 12 beats per second, to spring when the rolls are faster. The black woodpecker *Dryocopos martius* of central and northern Europe, incidentally, produces a drum roll of 40 beats, which lasts for a couple of seconds, and black woodpeckers in different geographical regions have drum rolls of different durations – distinct drumming dialects.

Drumming appears to be important during courtship, to bring the female into the breeding condition, and for territorial proclamation. Playback sounds of drumming will elicit drumming behaviour, for instance, in male great spotted woodpeckers. Wallschläger's work has also revealed that, contained in the drumming pattern, are factors which show individual identity. Each woodpecker has its own drum pattern, so a male and female can instantly recognise one another, as well as their neighbours, by sound.

Drumming is thought to have evolved as a byproduct of feeding behaviour. Woodpeckers attack the trunk of the tree for beetle larvae and the like, hammering their beaks into cracks and crevices. Green woodpeckers *Picus viridis*, unlike their cousins, do not drum so much, but give a loud laugh-like call known as the 'yaffle'.

Bird Songs

Usually it is the male bird that sings. There are a few exceptions, but in about 95% of cases the male is the solo songster. The male only sings at certain times of the year. In Europe, most song birds sing in spring. The song is often more complicated than a call, and naturalists over the years have associated this more complex signal with the transmission of a more complicated message.

The non-evolutionists, non-selectionists or fundamentalists have suggested that bird song has no function, and that birds make their music for their own pleasure, or indeed, as some have proposed, for the pleasure of listening ornithologists. Romantic as this may seem, there is unfortunately no evidence that birds produce their songs for the sheer joy of it. A song is difficult to produce, expensive in energy terms (as energy-consuming as flying, according to some estimates) and can be very dangerous in exposing the individual to predators. A small wren, for example, belts out a very loud song for its size; the entire body shakes and quivers. In nature, it is unlikely that a bird would make such an effort, consume so much energy which requires so much food, if it did not have a good reason for doing so.

At present there is no clear evidence that bird songs contain more complex messages than calls. There appears to be no complicated code or language which we might unravel. Bird song, simply, conveys the information more effectively; the message is getting across better. One feature that is immediately obvious is repetition; the same syllables and notes are repeated in the same order and in the same pattern time and time again. With a background of the songs and calls of other animals, atmospheric noises, and other extraneous sounds in the environment, a bird might find it difficult to get its message through. By repeating its signal a bird will eventually get its message to the intended recipient, with less risk of it being lost in the transmission process.

Male birds sing either to repel other males or to attract females. Charles Darwin proposed that a male song bird's main interest is in charming a female to approach, share its territory, and be its mate, and that mate attraction is the principal function of bird song. Eliot Howard, on the other hand, writing in *Territory in Bird Life* in 1920, convinced most naturalists that bird song is mainly concerned with staking out and defending a territory. Which, then, is the more fundamental function of song?

It could be that the same song has two different meanings, depending on which individual is receiving it. A resident male might sing to repel an intruding *male* that enters its territory, yet sing the same song to attract a passing *female*. It is not uncommon for song to have this duality. Any sophisticated communication signal is coloured by the context in which it is given and received, so the way these functions emerge depends on the life cycle of the bird. In the temperate regions of Europe or North America the male birds tend to arrive at the breeding areas first. They go through a phase of intensive competition for territories but, after two weeks at the most, they will have staked their claims and be sitting pretty on the basis of whatever agreement has been reached with neighbours. They don't stop singing; there is simply a shift in function. Singing continues, but to attract and stimulate females. So, even though the same song may be used for both territorial proclamation and courtship, there is often a separation in time.

Not all birds, though, show this duality of song function. Song may be weighted more towards one than the other. Peter Marler has suggested there is a continuum of species, from those in which mate attraction is the more important function, through to those which are more concerned with the repulsion of rivals, with all shades of grey in between.

If the sole aim is to attract a female, a male's song should be varied, with no breaks, and terminated after a successful mating. When repelling rival males, a male's song can be simple and should be spaced out in order to listen for replies. Charles Hartshorne, in 1956, described some birds as 'continuous' singers – males produce a stream of song with no gaps – and others as 'discontinuous' singers – they sing for a few seconds, pause, sing again, and so on. Hartshorne, originally, interpreted this finding in his 'monotony threshold hypothesis' – put simply, the more continuously a bird sang, and the greater the repertoire, the less 'bored' a listener would be. More recently, Peter Slater, of the University of Sussex, has pointed out that, if a song is to be used in a two-way conversation, such as that between territorial males, then there have to be gaps in the performance so that each may hear the other's reply. 'Discontinuous' singers, Slater argues, are likely to be species that use songs in territorial interactions between rival males. They would tend to copy very accurately in order to match their neighbours, for their language has to be shared. 'Discontinuous' singing, in addition, has the advantage that gaps between or within songs allow the bird to keep an eye and ear open for predators, especially as it is drawing so much attention to itself in a vocal border dispute.

'Continuous' singers do not wait for a reply. They are interested mainly in attracting a mate. Continuous singing is likely to result in more improvisation and less copying. The song is simply a continuous announcement that an unmated male is here, ready, and waiting.

The great tit, chaffinch, and white-crowned sparrow *Zonotrichia leucophrys* are extreme examples of 'discontinuous' singers, and the nightingale *Luscinia megarhynchos*, sedge warbler *Acrocephalus schoenobaenus*, grasshopper warbler *Locustella naevia* and brown thrasher *Toxostoma rufum*, are 'continuous' singers.

Song may have other functions. It is important that the receptive female is raised to the same state of readiness to mate as the male if mating is to be successful and fertilisation to take place. At Occidental College, Luis Baptista and Martin Morton carried out tests to see if the singing of male white-crowned sparrows in any way influenced the physiology of the female. During the winter, Baptista and Morton created artificial spring and summer conditions in the laboratory. Because it was winter, and out of the normal breeding season, the females were not physiologically ready to mate. In the artificial spring with extended daylength, however, the female ovaries grew. In one group of birds tape recordings of males were played and the ovarian growth rates were considerably higher than in those that heard nothing.

In some colonial birds, courtship songs and calls may bring an entire colony into synchrony, the advantage being that all the eggs and young will arrive and grow up together; there is safety in numbers.

Some birds sing 'prettier' songs than others. The male blackbird, for example, sings an elaborate song with a variety of song types, repetition of phrases and variations on song themes. But other birds can achieve the same results with a much less complicated signal. So, why the elaboration? Is it that the basic information, which repels male rivals and attracts females, is transmitted more effectively in a complicated song, or is there more information contained within a signal than we have so far been able to identify?

Clearly, redundancy, that is the repeated transmission of the same signal, will get the message through, eventually, to the listener, despite background noise and other environmental constraints. But that still does not account for the beautiful complexity of the melody of a blackbird's song.

To look for an answer we need to turn to Darwin and evolutionary theory. If the ultimate function of bird song, whether through territory acquisition or courtship, is to facilitate successful breeding, then sexual selection would operate most strongly when there is intensive competition for mates. Song elaboration may be an example of this kind of competition. Donald Kroodsma, of the University of Massachusetts at Amherst, wanted to know why some male birds have more than one song type. He took 24 unmated, virgin canaries and divided them into two groups. To one group he played the normal complex songs with 35 song types. To the other he played artificially simplified songs containing as few as five song types. It turned out that the female canaries exposed to the more complex songs were turned on physiologically to breed more quickly than were the females that heard the simple songs. He was able to monitor this by using a technique developed by Robert Hinde at Cambridge. Bundles of ten-centimetre long strings were placed in the cages, and each female would place the strings into a nest cup as if building a nest. Kroodsma simply had to count daily the number of strings being used. Those hearing the complex songs built their nests faster. They also tended to lay more eggs, although without a male present the eggs were infertile. This suggested that males with larger repertoires, within a species, might have some kind of advantage when the females choose their mates. If the male can 'impress' the female with a more complex song and larger repertoire, then it might say something about the quality of his breeding potential. From the work of Fernando Nottebohm it is known that male canaries with larger song repertoires have larger song control centres in the brain. For a male song bird to learn and develop a large song repertoire must mean a considerable investment in energy terms. Maybe the female recognises this investment. In the canary, males tend to learn more songs and develop larger repertoires as they get older. Maybe the female can spot the correlation between the number of different songs and the age of the male, and choose the older, more mature and successful bird. It is known in, for instance, the redwinged blackbird *Agelaius phoeniceus* that the older birds are the more successful breeders. The same is true of some other species of birds and of frogs and toads.

In another study, this time in the wild and with Bewick's wrens *Thyromanes bewickii*, Donald Kroodsma followed the movements and activities of birds identified by coloured bands on their legs. The wren population in Oregon sings throughout the year. He sat, watched, and recorded the songs of seven individuals, from one or two o'clock in the morning until way after dawn, usually until about 11 o'clock. He found that when the number of songs each male sang was plotted against the date it had hatched the previous year there was an interesting correlation; the earlier the male had hatched, the larger was the number of songs in its repertoire the following year. Kroodsma suggested that the female can recognise this. After all, a male hatched earlier in the year might have a better choice of territories than a male hatched later, and a better territory would give the female a better chance of reproducing successfully. Kroodsma emphasises that the reasoning is a little contorted and requires several inferential leaps. However, it seems likely that the sooner a male starts learning, gains a territory, attracts a mate, and begins to breed the greater chance he has of gaining an advantage over his rivals, and this may be associated with the quality of his song.

For some species, therefore, a large repertoire appears to be desirable. By having a rich and varied repertoire, a bird is thought to gain some evolutionary advantage. It came, then, as a surprise for Donald Kroodsma, in collaboration with Rick Kennedy at Rockefeller University, to find that birds of the same species in different habitats might not have the same aspirations in repertoire size. They found, for example, east-west differences in the songs of winter wrens, *Troglodytes troglodytes hiemalis.* In Oregon, males may have up to 30 different songs, while in New York and around Massachusetts they have only two. The European birds have three or four different songs. The same thing has been observed in long-billed marsh wrens *Cistothorus palustris.* Western males have an average of 150 different songs, eastern males have about 50. This is correlated with a reduction in the number of song control centres in the brain of the eastern birds. The question with which the researchers are left is whether the difference is environmentally induced or genetic. Kroodsma awaits the results of new experiments.

Whether 'continuous' or 'discontinuous', repulsive, attractive or stimulating, a male song bird must get started with his performance; and having started must know when to stop. What triggers a bird to sing, and how does it know when to terminate its song?

If a male bird is isolated in a soundproof chamber so that he can hear nothing other than the sounds of his own movements, he will sing to a regular schedule. He will commence singing early in the morning as if in the wild and part of the dawn chorus. There is an endogenous circadian rhythm, an internal clock, that governs the times he sings and guides his singing programme through his repertoire of songs. Superimposed on that schedule, in the wild, are other stimuli that might trigger song production. If a rival male starts up, for instance, then a male will sing in reply.

Inhibition of singing might well result from the sudden arrival of a female. The male, in this case, may change over to another set of vocalisations which are involved in close-range courtship. In some birds, at that moment, song virtually ceases for the rest of the season. A 'continuous' singing bird still singing late in the season is certain to be one of the few that has failed to obtain a mate.

The Dawn Chorus

At the break of day and at dusk, many woodland birds indulge in long bouts of singing and countersinging. 'To rise with the lark' or at 'cockcrow', we often say; but why should the lark and the cock be calling at that hour? Singing, as just one of a bird's daily activities, is clearly competing in a bird's schedule with foraging for food, or keeping watch for predators. Why, then, should singing be the dominant activity during the early part of the morning?

To find an answer to this, John Krebs and Alex Kacelnik, of Oxford University, looked at the way great tits apportion their time to different activities throughout the day. They reasoned that early in the morning, when the bird wakes up, the day is dark and cold and not a particularly good time for food gathering. Food availability is generally fairly low. The insects, on which great tits feed, would be inactive and difficult to locate in the half-light. With little food around, the motivation to feed would be low. John Krebs had also noticed, from his field observations, that many birds are actively seeking living space early in the day, so there would be high pressure on a territory holder from wandering birds. Both these factors, if they could be shown to operate, would encourage birds to sing at dawn.

To test this hypothesis, Alex Kacelnik set up large aviaries and introduced male great tits who promptly established territories. Kacelnik was then able to alter several factors – for example, the availability of food and the pressure from rivals. A third factor he looked at was the bird's internal clock. Might a daily clock, independent of food or rivals, govern when the bird sings or when to feed? By looking at these three factors, Kacelnik was able to establish that food was the major influence determining whether a bird sings at dawn or not. A bird presented with an abundance of food early in the morning is less likely to give up feeding in order to sing at, and chase away, a rival. If the food is reduced in quantity there is a greater chance the bird will sing.

Kacelnik also measured the efficiency with which birds feed at low light intensities and low temperatures and confirmed what Krebs had intuitively expected; that given the temperatures and light intensities normally found at dawn, birds are less good at feeding than, say, three hours after dawn. With atmospheric conditions unfavourable for foraging, there is likely to be greater pressure from territory-seeking males. Dawn, therefore, is the part of the day when the greatest benefit is derived from chasing them away.

Acoustic theoreticians had already shown that transmissions are more effective at dawn because air turbulence is at a minimum and temperature gradients are favourable. But the work of Krebs and Kacelnik revealed that the picture is far more complex than was first thought, and that food availability and pressure from intruders play a major part in any explanation of why birds sing at dawn.

Territorial Proclamation

Song can be a long range 'keep out' signal, a proclamation of ownership, but how do we know that there are not other forms of signal that are just as important?

In order to check this, John Krebs went back to square one. As a population ecologist, his interest in bird song came from the population dynamics of birds. For his thesis, Krebs was investigating whether the breeding density of birds in a particular habitat is limited by territorial behaviour. First he wanted to be sure that intruders are kept out by some kind of behaviour on the part of the territory holder. One spring, he carried out some simple experiments which involved removing territory holders from a breeding population of great tits. He found that new birds would enter the vacated territories, filling up the empty spaces. The new arrivals, he noticed, came from poor habitats nearby, where breeding success was low. They had been waiting on the sidelines, ready to move house when a better living space became available. How do the birds know whether a territory is occupied or not? The intriguing thing was that the birds would come in very quickly after the removal of the territory holder. The obvious answer seemed to be that they listened to see if a resident was singing. Sound signals would be more effective in a forest than, say, visual or olfactory cues.

Krebs then checked the literature for references to song and territorial proclamation. There were lots of studies showing that, when song was played back to a territory holder, it would react by approaching and maybe attacking the loudspeaker, and this was taken as evidence that song acts as a territorial signal. But no-one had checked that, when a bird sings in its territory, it is actually saying 'keep out', and therefore defending its territory against the birds waiting in the wings. In 1975 Krebs carried out his classic, yet simple, removal experiment. He took away territory holders, as before, leaving empty spaces in the wood. This time he replaced some of them with loudspeakers playing the songs of the birds he'd taken away. In other territories he put loudspeakers playing a control sound – a tune played on a tin-whistle, which had notes at about the right frequency and duration. Other territories were left empty. He sat and watched. Sure enough, in came the birds from outside the study area to occupy the empty spaces. The territories with normal song were treated as if a bird was in residence and were avoided. Instead, the new arrivals went first to the tin-whistle or silent spaces. This seemed to show that the initial idea was correct – male great tits waiting to enter and take over territories know whether a territory is occupied or not simply by listening for the resident male's song.

Having established a territory, a bird must maintain and defend its borders. In some species, neighbours sing at each other across the line of demarcation. Often they sing the same song, a form of countersinging known as 'matching'. But which bird should sing first? And, as most birds have more than one song type, how does an individual, having started, know which song type to sing next? Donald Kroodsma has uncovered some clues while working with the long-billed marsh wren, a bird with a large, rapidly sung, song repertoire. He was fascinated by the way males in neighbouring territories sing the same song repertoire back and forth to one another each day. Bird species with very simple songs would have little scope for improvisation when matching with neighbours. But with a couple of male long-billed marsh wrens, with over a hundred song types each, the number of possible ways the birds could sing

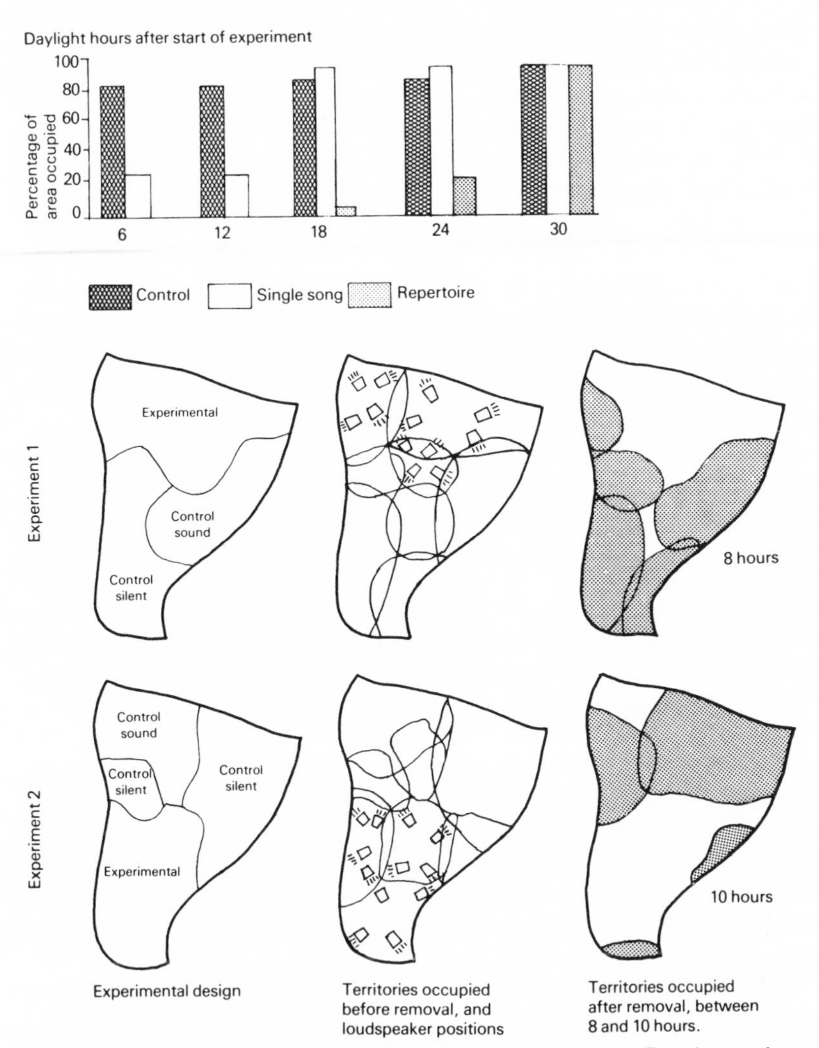

Fig. 18 Removal and playback experiments with great tit songs. Resident males are removed and loudspeakers placed in certain territories.

together is limitless. How do they decide the order in which to sing? In a pilot project Kroodsma taught two wrens nine song types. They were allowed to sing at each other quite freely and it soon became apparent that the physically larger male became dominant, and led the singing. The physically submissive bird would match the other's song and follow the dominant bird's song-singing order. Kroodsma then interfered with the singing by increasing the loudness of the submissive bird's singing. The dominant bird began to follow the amplified bird's song and the previously submissive bird started to lead the proceedings.

In the field, Jerrard Verner studied eastern Washington long-billed marsh wrens. He placed a loudspeaker playing marsh wren songs in an established territory to simulate an intruding male. Normally a resident bird is dominant to an intruder, and Verner found that, because the order of singing in marsh wrens is so stereotyped, the resident bird was able to anticipate the order of songs on the tape recordings and get in with the matched songs first. In this way it actually led the tape recorder! Why the submissive male should play this game is not clear, although Kroodsma has some hunches. Territoriality might be viewed, for example, as an exploded dominance hierarchy. Males entering a suitable breeding area must toe the line before they are able to establish themselves. By matching the songs of dominant residents they might gain a foothold. And maybe females searching for a mate are more responsive to the bird in the leading role – the bird higher in the 'pecking' order.

But, if matching is taking place, to what standard are the songs being compared? Peter Marler thinks an individual's own song is important. W. H. Thorpe, and later Fernando Nottebohm, developed the hypothesis that some of the imagery a bird has of its species song may develop around its experience of its own individual song. If a male memorises its own singing , uses those memory traces as a basis for matching the songs of others, and only accepts as rivals those that are a close match to its mental picture, then playback of its own song should have interesting results, because that is going to be a perfect match to its image of what the species song should be like.

In some experiments, a male's own song has turned out to have an intermediate rank as a stimulus, falling somewhere between the song of a familiar neighbour and the song of a stranger. But still there is evidence that a male's own song has some unique status as a sound heard, but just what its role may be in development is not clear. This unresolved issue is of interest at present because of the notion of 'self'. In human development, for instance, a child is considered to reach social maturity when it develops some sense of its own personal identity, and can use that as a basis for relating to others. The search is on for a similar phenomenon in other animals.

John Krebs, together with Bruce Falls of Toronto University, looked at matching in the great tits of Wytham Wood. Their starting point was to play to birds songs from their own repertoire and record whether or not they were matched. They found, not surprisingly, that great tits match their own songs. What they also found is that the reliability with which they matched varied according to circumstances, but it was not clear what those circumstances were. They looked in more detail.

In another experiment birds were presented with three different kinds of playback of song types from their own repertoires: firstly their own version of that song type; secondly a neighbour's version; and thirdly a version of the same song type sung by a stranger. Typically, birds would approach the loudspeaker, fly about looking for an opponent, and then begin countersinging. The intensity of the challenge was not always the same, though. A neighbour's version of the song resulted in a weak response. The resident bird would approach slowly and sing little. A stranger's version would evoke a full-bodied challenge which sometimes ended with the resident attacking the loudspeaker. Bruce Falls feels that this may be because the stranger poses a greater threat than does the neighbour. The neighbour has a territory of its own and is recognised by the resident bird which does not get too excited. This reduces the amount of unnecessary effort, time and risk spent in fighting and chasing, by ignoring, to a certain extent, the songs of immediate neighbours.

The accuracy of matching varied too. Birds most reliably matched their own songs; next most reliably matched the songs of neighbours, and least reliably matched the songs of strangers. This still did not tell what purpose matching serves; rather it showed that matching is based on a precise similarity between the song version played to the bird and the bird's own rendition of that particularly song-type. Krebs looked at this data again and came up with a possible functional significance of matching.

Krebs remembered the environmental degradation experiments with Carolina wrens carried out by Haven Wiley and Douglas Richards. They had revealed that Carolina wrens are able to judge the distance to a rival bird, which is singing, simply by listening to the way the signal is degraded by echoes, reverberations and other atmospheric disturbances. But how, wondered Krebs, can a bird actually tell if a song is degraded? It would need to compare the song it hears with a standard of what the song would be like if it were not degraded. Krebs concluded that the standard is in its own repertoire. When a resident hears a song, it is compared with its own rendition of the same song and the bird can assess whether the song is degraded or normal, thus discovering the distance of a rival. So, if two neighbours match each other's song types, the receiver can say 'I have the same song type as you, therefore I can judge exactly how far away you are, because I can measure the degradation of your song', and the original singer is telling how far away it is. It is a mutual exchange of information about distance, and reliable too. The degradation depends solely on the properties of the habitat, preventing the birds from cheating on the system by pretending to be closer or further away than they really are.

By matching to its own song, a bird can tell exactly whether a neighbour is, say, 25 metres and not 20 metres away. The five metres difference could be important because a neighbour 25 metres away might be inside its own territory and no threat, but 20 metres away it might be inside the other's patch and attempting to steal a piece of ground. Neighbours save time and energy if they keep telling each other how far apart they are.

If the song a bird hears is very different from its own version it would be difficult to judge easily the amount of degradation; hence the more violent or alarmed reactions noted with intruding strangers.

Repertoire size was another topic which interested Krebs. The well-known argument to which he turned his attention was that if bird song is saying something fairly simply like 'keep out, this is my territory', or 'here I am an unmated male', why is it said in such an elaborate and complicated way? Krebs went back to his removal experiments, taking male great tits out of their territories and replacing them with loudspeakers. The playback experiments were repeated but this time different numbers of songs were played in the territories with loudspeakers. In one territory a varied repertoire of song types was played, while in another a single song type was played over and over again. Krebs reasoned that if song does play a role as a 'keep out' signal, then perhaps the repertoire, in some way, increases the effectiveness of song in repelling rivals.

When the residents were removed and the tapes played back, new birds would arrive and scout around to find the empty spaces. The spaces with no sounds were quickly reoccupied as before, but interestingly the spaces playing the single repetitive song type were more prone to occupation than those with the varied repertoire. Larger repertoires, it seems, make better 'keep out' signals; but why? John Krebs feels the mechanism is a very simple biological property of animal nervous systems. When exposed to a repetitive stimulus an animal often habituates, that is, it stops responding to the stimulus. If a territory holder sings the same song over and over again, rivals might gradually get used to it. If the resident bird changes its tune, the variety in its song reduces the chance that others will habituate to it. Why wandering birds should habituate to song is not at all clear. Maybe it is simply a property of their brains. Krebs would prefer to find a functional explanation, but as yet there isn't one.

Territory holders with varied repertoires have an opportunity to cheat. Wandering birds will be on the lookout for suitable areas with few resident birds, where there is less competition for mates and food resources. By singing a variety of songs, a resident bird might give the false impression of there being many other birds in the area. The deceit could be further enhanced by constantly moving about the territory singing from song-post to song-post. It would seem more profitable, then, for an intruder to look elsewhere. This has been named the Beau Geste Hypothesis after the novel by P. C. Wren in which the gallant legionnaire successfully defended a desert fort by propping up dead bodies around the ramparts to make it look as though there were many more defenders.

One bird which is seen to dash about its territory incessantly – often to the irritation of the wildlife sound recordist – is Cetti's warbler *Cettia cetti.* It may sing its first song in front of you, but it will disappear for a few seconds only to reappear at another song-post, a little further away, to sing its next song. Recent research by Ken Yasukawa, while at the Rockefeller University Field Station at Millbrook, with the red-winged blackbird of North America has given some credence to the Beau Geste Hypothesis. Analysis of song length and the frequency with which songs are sung has shown that they are as variable within a single bird's repertoire as they are between repertoires of a group of different birds. Also, male red-winged blackbirds tend to change song types when they change song-posts.

So, come foul means or fair, if a song repertoire is important in obtaining the best territories, there should be a link between repertoire size and reproductive success. Male great tits, for instance, breeding in the choice territories should have mates laying the largest clutches of eggs and therefore producing the greatest number of surviving young to breed the following year. In Wytham Wood, John Krebs, Chris Perrins and Peter MacGregor recorded the songs of male great tits and compared individual song repertoires with the ability to rear successful families. As part of a long-term population study of the great tit, Chris Perrins and his assistants go around the wood and count the number of eggs laid by each pair of birds, how many hatch out to produce young, and the number of young that survive through the winter to breed the next season. Peter MacGregor and John Krebs could thus record the songs of known individual males.

The Oxford researchers showed that there was, indeed, a relationship between the size of song repertoire and breeding success, although the result was not the one expected. Males with poor repertoires containing one or two song types, and males with large repertoires of seven or eight did not fare so well as those with middle-sized repertoires of three to four song types. There seems to be, in great tits at least, an advantage in having a certain number of variants but a disadvantage in having too many. Males with poor repertoires do worse in terms of reproductive success than males with several song types. So, although John Krebs is comforted to find that song variants do have enhancing effects on the territorial keep-out signal, which are reflected in reproductive success, he is at a loss to explain the disadvantage of very large repertoires.

Courtship

Human observers are often attracted by the conspicuous and repetitive movements and the loud elaborate sounds made by some birds prior to mating. Courtship is an important aspect of bird behaviour, for it creates the right conditions in which the male and female of the species can come together in order to reproduce. For birds, visual and acoustic signals are used as important sexual attractants. For birds in marsh or forest, where visibility is restricted, songs and calls are often the only means of long-distance communication. In song birds, it is the male that sings. Instead of pursuing a female, a male sedge warbler, for example, sits and sings in his territory and waits for the female to come of her own accord, attracted by his song. The problem is the sound signal might not only attract a mate but also a predator. A courtship display is a compromise between attracting a member of the opposite sex and exposure to a predator. And, to complicate things still further, to avoid wasted effort, courtship singing must attract only females of the same species. Song must be recognisably distinct for each species. A bird is also likely to be agitated about the close proximity of another individual of the same species. The potential mate could be confused with a marauding intruder or a competitor for food. A male's song, therefore, must not only attract the female, it must also suppress her natural tendency to flee.

From the female's point of view, not all males are equally suitable for mating. She must have a means of selecting the fittest. The strongest and healthiest males usually have the highest reproductive potential – they will look after the brood better, and have the best territory with the most food. These are short-term benefits. In the long term the male offspring are likely to inherit the father's ability to maintain and defend the best available territories, so there is an obvious genetic advantage in selecting the best available male. A male's song may indicate to a female his suitability as a mate. So could it be that the ultimate function of bird song is sexual attraction? Clive Catchpole, of Bedford College, University of London, thinks this may be so. He has been working with several species of warblers which breed in Europe. They are the *Acrocephalus* warblers, including the reed warbler *A. scirpaceus* and sedge warbler *A. schoenobaenus* which breed commonly in Britain, and the marsh warbler *A. palustris* which is largely confined in Britain to the south-west of England.

The *Acrocephalus* warblers are small, well camouflaged birds which live in dense marshland vegetation. As they cannot see each other clearly there has been an obvious premium on vocal, rather than visual communication in their evolution. The first clue Catchpole was able to uncover that song might be particularly important in sexual attraction for these species came from studies on their seasonal and daily rhythms of song production. They are migratory birds, the males arriving in Europe during the spring to take up territories in which they sing for long periods of time, sometimes throughout the day and night. The females arrive a few days later, find a mate, and settle in for the breeding season.

After pair formation the male's routine changes – he reduces song production, and in the case of the sedge warbler, stops singing altogether. This indicated to Clive Catchpole that the male is using his song to attract a female. Once he has found her there is no need to sing, so he stops. Catchpole was able to confirm this with playback experiments. He placed a loudspeaker in a sedge warbler's territory and played a rival's song. The resident would approach the source of the sound aggressively, hunt around, fluff-up in a threat display, but he would not sing. He would still defend his territory, but not with song.

Further evidence for sexual attraction, suggests Catchpole, lies in the warbler's elaborate song. The sedge warbler can have a hundred or more separate syllables. The male selects from his repertoire several syllables which are composed into a long and elaborate song by repeating them, alternating them, and recombining them in a variety of patterns. The patterns are so

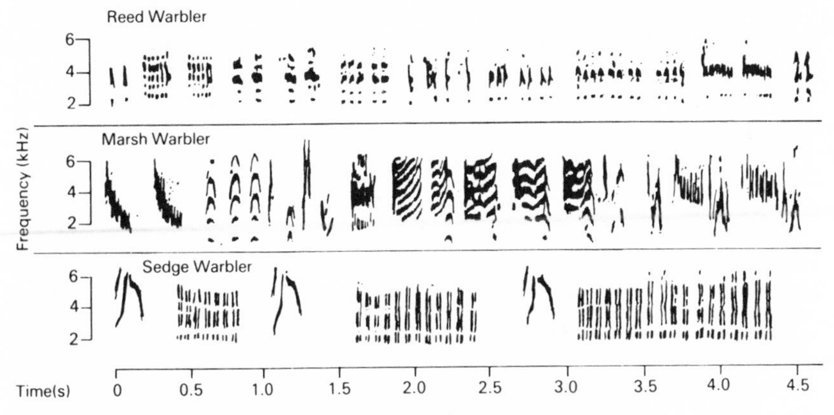

Fig. 19 Sonagrams of the songs of three *Acrocephalus* warblers.

variable that Catchpole, having analysed miles of recorded tape, has been unable to find an instance of a sedge warbler repeating a song. Each song is a separate musical composition, a distinctive behavioural event. Catchpole likens the sedge warbler song to the peacock's tail in that it is a flamboyant and extravagant signal used for impressing and attracting females. To test whether male sedge warblers singing complicated songs have greater success in attracting females than do those with simpler songs, Catchpole recorded a wild population of sedge warblers in England. He noted when each bird arrived, and if and when each one attracted a mate. This was relatively easy to do because mated sedge warblers stop singing. So, the date of arrival and the date on which a male attracted a female to his territory was filed away with a tape recording of his song. The songs were then analysed with the sound spectrograph and the complexity of each rendition assessed. The number of elements in each bird's repertoire was plotted against the pairing date and it was revealed that the birds with the most complicated songs attracted females first. Those with simple song repertoires had to wait. Birds with elaborate songs clearly had some advantage over their rivals.

Not all species of warblers produce such complex songs as the sedge warbler, and yet others sing with even more intricate song patterns. By looking at the species in his study group, Clive Catchpole sought some clue to the evolution of the songs. He looked at their ecology and behaviour and found major differences.

Some *Acrocephalus* warblers are monogamous – one female per male – and others are polygamous – each male attracting several females. Most bird species are monogamous. Both parents are needed to bring food to the nestlings or the chicks may starve. However, in certain habitats food can be plentiful. In marshland, for example, insect life can be super-abundant. In areas rich with food supplies, both parents may not be needed to feed the young. A female could catch enough large insects near the nest to bring up the youngsters by herself. The male, in that circumstance, might desert the female confident that his genetic investment is safe, and polygamy evolves. In the *Acrocephalus* warblers, polygamy has evolved in two European species, the aquatic warbler *A. paludicola* and the great reed warbler *A. arundinaceus.*

Clive Catchpole first reasoned that, if sexual selection has driven the evolution of elaborate bird song, then polygamous males will have evolved more elaborate songs. A 'showy' male, attractive to a bevy of females, would leave behind more offspring. Catchpole tested the prediction by assessing the song syllable repertoire of each species. He took the results from the six European warblers, but was in for a surprise. It was, in fact, the polygamous species, the aquatic and great reed warbler, that had a song repertoire of only ten to 20 syllables, and the monogamous species, the reed warbler, the sedge warbler, the marsh warbler, and the moustached warbler *A. melanopogon* that had the most complicated songs with 35–100 song syllables. The polygamous bird sang short simple songs while the monogamous males produced long, continuous and variable bouts of song. For an explanation Catchpole looked at the birds' ecology and behaviour.

Species	*Repertoire range: Number of syllables*	*Mean Length of song in seconds*	*Mating system*
Marsh warbler *(A. palustris)*	80-100	Continuous	Monogamous
Reed warbler *(A. scirpaceus)*	70-90	Continuous	Monogamous
Moustached warbler *(A. melanopogon)*	60-80	Continuous	Monogamous
Sedge warbler *(A. schoenobaenus)*	35-55	19.49	Monogamous
Great reed warbler *(A. arundinaceus)*	10-20	3.20	Polygynous
Aquatic warbler *(A. paludicola)*	10-20	2.18	Polygynous

Fig. 20 Repertoire ranges, song lengths and mating systems of six European *Acrocephalus* warblers.

It seemed to Catchpole that, in a polygamous species, a female would be less concerned with the quality of the male, as she is likely to be deserted and left to bring up the chicks herself. To maximise her reproductive success she should be more concerned with territory and the food it will provide for her young. The female is not selecting her mate directly on his own attributes, but indirectly through the quality of the territory. Polygamous species defend large territories in rich marshland. So male song, Catchpole believes, may have evolved primarily within the context of competition between males for territories – intra-sexual-selection pressure – with the resulting short, simple songs. This further supports Peter Slater's ideas on 'continuous' and 'discontinuous' singers.

In monogamous species, with long elaborate songs, the male is important in helping to feed the young. The female chooses a good quality male rather than a good quality territory. Territories in monogamous species of *Acrocephalus* tend to be small, sometimes in drier areas, with most food being obtained from way outside the territorial boundaries. The birds make long flights in search of food supplies so the male must be strong, physically capable of many arduous flights for foraging. The first indication a female may have of a male's suitability is his song which she hears some distance away. She may, or may not be attracted by what she hears. This is inter-sexual selection at work. As in the sedge warbler study, the males with the most elaborate songs obtain their mates first.

There may in fact be a mutual advantage – the male, by singing an elaborate song, reaches the more active and alert female, while the early female gets to choose the cream of the males.

Clive Catchpole believes that the ultimate function of bird song is to obtain a female, breed successfully and produce more offspring than other individuals. Even those birds singing to defend territories are selected for through the quality of the living space. The end result is that a male obtains a female by the indirect route of proclaiming a territory. Catchpole also points out that the repulsion of males and the attraction of females are not mutually exclusive. A song may serve both functions, although in different proportions. In the sedge warbler, which stops singing after pairing, it is clearly mate attraction that is more important. In other species, which reduce their song output after pairing, but retain it for use later, it may well be that there is some territorial function in the song. The great reed warbler has two functional versions of its song. It is a simple song, but with a certain amount of variation. It starts with a number of short clicking sounds and expands into more melodious notes towards the end. When the bird is setting up a territory and trying to attract a mate it sings the longer, more elaborate version of the song, each song lasting from between three to 20 seconds. When a female has been attracted, the male stops singing the long complicated song and instead sings the clicking sounds. Catchpole confirmed this with playback experiments. He placed a loudspeaker in the territories of birds which were singing the long more elaborate song. If he played back the song of a rival, the resident bird would approach the loudspeaker, at the same time shortening his song to the simple clicks. So, the great reed warbler sings two kinds of songs – the longer, complicated version used to attract a female, and the shorter, simpler, more stereotyped version it continues to use to keep out rival males.

6
SONG LEARNING

How does a songbird acquire its song? Is it inherited or learned, nature or nurture? Some of the first attempts to answer this question were made by Professor W. H. Thorpe at Cambridge University. He was one of the first ethologists to use the sound spectrograph, which had been developed for analysing human speech. It was with the advent of this machine that a whole new window was opened on the study of animal sound communication.

A simple way of checking for evidence of learning is to isolate hand-reared birds from any sounds that they might learn. Thorpe's early work was with chaffinches. He found that birds kept in soundproof boxes were unable to sing properly. They would attempt to sing normal song, but unless brought up in the company of other normal birds, they could only attempt rather raucous noises quite unlike normal song. Male song birds sang abnormal songs if deprived of an opportunity to hear adult song during the first nine months of life. But, if played recorded tapes of chaffinch song, they would learn to sing accurately as early as ten days after hatching. It became clear that songs are learned. At first Thorpe thought the learning process was progressive throughout the first year of a bird's life. It wasn't until later that Peter Slater and his colleagues at Sussex University demonstrated that fledglings could learn whole songs, thereby narrowing down the likely learning period.

Professor Peter Marler, now at Rockefeller University, New York, first worked with Thorpe's chaffinches at Cambridge. He then went to the University of California at Berkeley and together with Mayukasu Konishi, now at the California Institute of Technology, Pasadena, repeated the isolation experiments, but this time with local white-crowned sparrows *Zonotrichia leucophrys* an especially common species in the San Francisco Bay area which shows very distinct local dialects. The white-crowned sparrow has just one song type, consisting of two parts. There are one or more introductory whistles followed by some brief, rapidly repeated syllables. Sometimes there is a whistle or buzz at the end. The trill portion contains the dialect differences, and with a little practice you can tell where you are in the Bay area just by stopping for a moment and listening to the nearest white-crowned sparrow.

Marler was able to confirm the chaffinch findings – white-crowned sparrows in social isolation developed abnormal songs. Although the song had the whistle sections present, the trill was absent. But there was a problem. Although the song developed by a male in social isolation is abnormal, in the sense that it sings something not heard in the field, there were still normal qualities to the song. Marler played these sounds to ornithologists and asked them to identify the species. Invariably they would suggest that the song must be a new species of *Zonotrichia.* Marler was puzzled – just what was the genesis of those normal components in the song that an isolated male is still able to develop? Konishi decided to look at the effect of deafening on song production, and established that the information enabling the socially isolated birds to show a hint of normal song lies somewhere in the auditory part of the brain. A deafened male develops a song which is even more remote from normality, more primitive than anything that an intact bird will develop, even in isolation. The song of an early deafened white-crowned sparrow is a raucous buzz, more like the sound of an insect than of a bird.

Marler and Konishi developed the results into a new interpretation of the song-learning process – the auditory template theory of song learning. They concluded that each song bird studied has some innate capacity to develop at least some of the normal features of the song as heard in the wild, and that that capacity rests on a mechanism associated with the sense of hearing. When a young male begins to sing, in the normal course of events, he listens to his own voice. He has an image of what it should sound like and he proceeds to match his singing to this expectation template fashion. He gradually achieves a more perfect match with the template until he is able to sing a song with some of the normal features.

In a bird hearing normal song, the template is modified, overridden by the learned information, until the song includes features of the local dialect. The male, now, has a new template which is more advanced than the innate version. He proceeds through this matching process, through sub-song (an immature form of song with impure notes delivered more quietly than full song and in longer phrases) and plastic song (the pre-adult song when phrases are shortened and notes perfected), until the song crystallises into an imitation of the local dialect to which the male was exposed when he was young.

White-crowned sparrows, chaffinches and other male song birds that have been studied seem to have a sensitive period for song learning, typically when they are quite young. In the white-crowned sparrow the period is somewhere between ten and 50 days, at a time when the bird has left the nest but is either dependent on or associated with its parents.

The young male white-crowned sparrow tends to move away from the area of hatching somewhere around two months of age. This, then, is a case where the youngster does his song learning while still at home, although he does not start to sing himself until some time later. When he does start to sing the following spring he generates an imitation of the local dialect. The imitation,

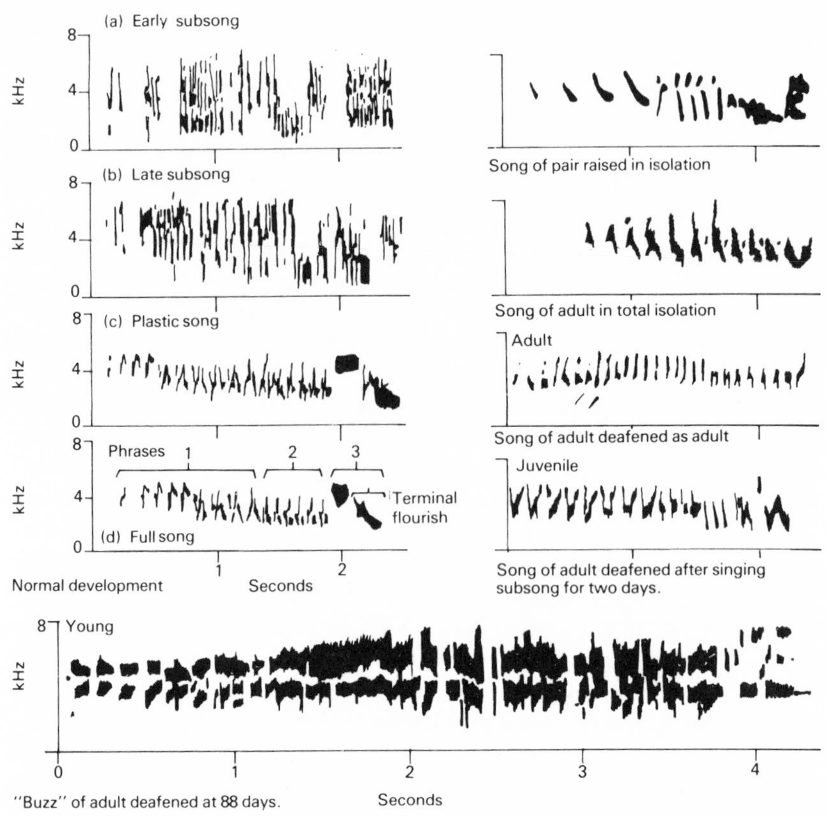

Fig. 21 Development of chaffinch song.

though, is not completely faithful to the models heard. A young male will extract certain features from those heard in infancy and keep them intact, while allowing himself freedom to improvise or invent other portions of the song. In the relatively simple song of the white-crowned sparrow there is a strong tendency to conform to local fashion, although each male includes individual, personalised features. Each male has a set of structural features that label him 'white-crowned sparrow'. They identify him as a member of a particular species. There are also features of local dialect which tell about the area in which he was brought up. In addition, there are portions, usually in the whistle section, which are his personal marks. These sound cues are employed in different ways in the social life of the birds so that they all know their immediate neighbours, can distinguish neighbours from more distant members of the community, can discriminate between dialects, and can identify members of their own species and other species. There is a hierarchy of song features which is present even in the relatively simple song of the white-crowned sparrow.

Sensitive periods for song learning are by no means universal. Those birds with well-defined dialects tend to have narrowly defined periods for learning; others have different learning patterns. In some, the programme of song learning remains open, sometimes for quite long periods and, in a few cases, throughout life. The red-winged blackbird continues to add to its repertoire as it matures, and the domestic form of the canary discards quite a large proportion of its song from one year to the next, adding new phrases to replace the old. In both these species the size of the song repertoire increases annually as the male gets older and more mature. Females choose the older, fitter males, which have survived several breeding seasons, in preference to the younger, untested juveniles.

In Wytham Wood, Oxfordshire, John Krebs and his colleagues, notably Peter MacGregor, also directed their attention to song learning periods. They wanted to find out at what time of year the young male great tits learn their songs and from whom. Traditionally it was thought that songs were learned from fathers. Laboratory studies with zebra finches *Poephila guttata* by Immelmann and with bullfinches *Pyrrhula pyrrhula* by Nicolai, both from West Germany, showed quite explicitly that caged male birds learn songs from the father. Krebs and MacGregor decided to look at birds in the wild to see whether young males sing like their fathers. Again, they used Chris Perrins' population of individually ringed great tits, many of whom are of known genealogy, and were able to look at the repertoires of fathers and of their offspring in the following year. The reason the Oxford researchers decided upon a field study is that laboratory studies do not offer a young male a choice of different song models from which to learn. The bird sits in a cage, perhaps with his parents or maybe a tutor tape. In the wild a young bird sits in the nest, or flies around after leaving the nest, and hears a variety of different songs. Does he know which song is his father's and then learn specifically his father's song or does he just pick up whatever he hears?

Looking at the songs of known fathers and sons, Krebs and MacGregor demonstrated that there is no tendency at all for young male great tits to learn songs from their fathers. Songs, in this species at least, are not passed from generation to generation within a family. So, from whom *does* he learn?

Further investigations revealed that the young male shares the highest proportion of his songs with his neighbours in the year that he sets up his first territory, that is, usually, when he is about nine months old. If the bird is hatched in June he will become a territorial male in March the following year. This suggests that, in great tits, quite a lot of song learning occurs in the first spring of life, rather than during the first few weeks as laboratory studies of other species would imply.

Not *all* the song elements in an individual's repertoire are learned from neighbours. In Wytham Wood some birds are 'resident', having been hatched in the wood, with known parents, and others are 'immigrants', having been raised outside the wood, which have wandered in from maybe several miles away. The elements of song repertoires are derived, seemingly, from the two groups. There are the 'common variants' which many different males in the wood include in their song repertoires, and the 'rare variants', which are sung by a few individuals in the population. The 'rare variants', it turns out, tend to be sung by immigrant birds which arrived as first-year males. Since the birds have predominantly 'Wytham Wood songs' but with the occasional song type from elsewhere, Krebs and MacGregor concluded that the local 'common' song types were learned on arrival at the wood and the 'rare' song types were learned before they dispersed from their home area.

So, from the field studies on great tits, Krebs and MacGregor have identified at least two periods when the birds can learn songs: some time early in life before they disperse at the age of two or three months, and some time later at six to nine months when they first set up territory. After the first year the repertoire is not modified. Krebs and MacGregor followed individuals for several years and found that they kept exactly the same song repertoire after their first spring.

Peter Slater, of Sussex University, working with captive chaffinches, has results which fit with the same conclusions. He has shown that hand-raised chaffinches can learn songs, both in their first few months of life, which corresponds to the pre-dispersal learning of wild birds, and in their first spring, corresponding to post-dispersal learning.

Krebs has also used the great tit data of songs of individuals with known family histories to find out whether the songs that a female hears early in her life influence her later behaviour. Female great tits, like many female song birds, don't sing very much, but that is not to say they do not know something about the characteristics of the song. In some species of song birds a female bird injected with male hormones will begin to sing a normal male song. The song is tucked away in the brain even if it is not normally sung. Do female great tits learn songs while still very young, particularly their father's song, and then use this information later on in life, when they pair with a male, to avoid an incestuous mating with the father? Chris Perrins' analysis of reproductive success has revealed that in-breeding in great tits is disadvantageous. Birds mating with close relatives are less successful in breeding than individuals mating with distant relatives or individuals from outside the population. Do the females, then, choose males with different songs from their fathers?

Krebs and MacGregor collected data for females where they knew the song repertoire of the female's father and the female's husband. Then they looked at the amount of sharing of songs between the two and compared the result with a chance expectation based on the rate of sharing of songs for the population as a whole. They found that females are less likely than expected by chance, to mate with males who share most of their repertoire with the female's father. In addition, they found that females are likely to turn down suitors where the song is too different from that of the father. Female great tits prefer to mate with males with song repertoires slightly different from the female's father, but not very different.

This result fits with the notion put forward by Pat Bateson of Cambridge University that females, in choosing a male with which to mate, avoid close genetic relatives and incest, but at the same time avoid out-breeding with genetically very distant males. Genetically very distant males may be adapted to different environmental conditions, so if there is any local population adaptation to different habitats, it would be advantageous for females to mate within their own population. The same is true for local dialects – different dialect groups might be adapted to different habitat conditions.

Between learning and the fully mature song is a period of great interest to researchers. This is when the learned song may change by identifiable stages, through sub-song to plastic song and finally adult song. Sub-song is rather like the babbling of a baby before it begins to talk, and as any mother knows it is a fascinating stage of development. But, it is also an almost impossible stage to understand or even document because the sounds are so variable. At Rockefeller University, Peter Marler, together with Susan Peters, undertook the daunting task of analysing sub-song, this time with song sparrows *Melospiza melodia* and swamp sparrows *M. georgiana.*

Like the white-crowned sparrow of California, swamp sparrows also learn from memory. They learn their songs when quite young; it is all over by 60–70 days of age. They do not come into full song until, perhaps, 200 days. Sub-song begins earlier, sometimes even when song learning itself is taking place at two months. It was thought traditionally that sub-song was a period of rehearsal, when songs are committed to memory as they are being heard, so that when the bird comes into full song the following spring the information is already firmly stored. Marler felt that sub-song is so unstructured that it was unlikely that rehearsal was taking place. He set out to analyse it.

A group of 16 birds was set up and recorded once a week for a year. Marler and his colleagues analysed the whole transition from sub-song, to plastic song, to adult song, and surveyed this enormous mass of material for evidence of rehearsal of recorded tapes to which the young birds had been exposed. As this was a laboratory study, Marler knew everything that the young birds had heard during infancy. He could look through the record for the very first appearance of anything that was remotely like the structure of the models that had been learned when the birds were young. Rehearsal, it turned out, did not begin until the birds were between 230 and 250 days of age. The male swamp sparrows had learned the tape recorded songs when they were young, committed them to memory by ear, without rehearsal, and stored them for up to 300 days before they began to produce anything remotely resembling the original model – a quite extraordinary feat. One of the targets of Marler's current research is to look at the form in which the birds commit the songs to memory.

Birds, Peter Marler suggests, have an ability to memorise a model when young, and break it up into sections, or syllables, which can be rearranged to create new songs. Birds, then, have the capacity to retain phonetic units that conform to the local fashion and yet can be recombined in new ways, allowing each male to place his personal mark on the song. They do this so freely that it is tempting to think that, when the song is committed to memory, it may well be done in a segmented form. By breaking the songs down into sub-units a bird could easily combine them later when it starts to sing in earnest. The units, interestingly enough, are not the smallest units of a song.

Birds, it seems, have a remarkable capacity to make use of something which is very much like the syntax of language. This is not the 'lexical syntax' of a human sentence; the components of bird song do not have independent meanings in the same way as human words. The song is a message as a whole. There is, however, another level of syntax called 'phonological syntax'. Combinations of consonants and vowels, or 'phonemes' from which words are constructed in human speech would be analogous to the syllables of bird song. For humans every language has a limited set of phonemes and an infinite number of ways in which they can be combined, giving an endless capacity to generate new words.

The simple song of a swamp sparrow has a dry repetitive trill, with a syllable repeated perhaps 20 times. Each of these syllables may be made up of six or seven distinct notes. The songs are broken down to the level of syllables and not notes. Marler feels this is, perhaps, a natural unit for the production of sound – the way in which some linguists view the phoneme.

In addition to the imitation, Marler's swamp sparrows also invented song material. As he had such a complete record of everything the birds had heard, Marler was able to detect these subtle variations. When the birds sang their full song the researchers would go through the spectrogram item by item, first looking for events attributable to learning in infancy. They discovered that some of the song components produced in adulthood did not match any of the models. The components were seen as copies at an intermediate stage of development but then they were changed. The bird had engaged in a kind of vocal play, imposing small transformations, and so getting further and further away from the original model. By the time the bird produces adult song the end product cannot be related easily to the starting point.

The red-winged blackbird is an obsessive 'inventor'. It is very difficult to identify imitation until the intermediate developmental stages are examined. Some birds generate seemingly completely original material that is not attributable, as far as can be seen, to the features of the original model.

In the original Cambridge chaffinch days the process of imitation was emphasised in song learning. The picture we now have is a lot more complicated. Along with imitation there is improvisation and invention going on. Birds are constantly generating new patterns of sound in nature.

One of the fascinating things Marler and his colleagues have discovered when analysing sub-song and plastic song in the swamp sparrow is that each bird generates far more song material in the course of development than is needed for normal adult song production. A male swamp sparrow has, typically, three song types. Marler analysed the plastic song a week before it had crystallised to adult song and found that the bird had been using six different syllables. A couple of weeks before it had been using 12 (in one case as many as 19). As it approaches the moment of song crystallisation, the song components get winnowed down very rapidly until the mature repertoire is left.

In a male producing 12 syllable types, Marler found that five or six were obvious imitations, one or two were improvised variations of an imitation, and four or five seemed to have been invented. Even though he had examined carefully the developmental record, he could not find ancestors in the learned song, nor were the invented syllables like the sounds produced by a socially isolated bird which has had no opportunity to learn. The syllables in 'isolate' song are simple, whereas the invented syllables are complex, although not as complex as the imitations. Marler concluded that the bird has a capacity for a truly creative process. He feels, though, that the creativity is not realised unless they are pushed over a certain threshold, as though their environment must in some way be enriched to a degree that motivates them to indulge in this creative activity. This is the target for future research – what kinds of conditions provoke creative invention?

The problem confronting Marler is not new. Much psychological research indicates that enriched environments are important to young organisms, not just to provide models for imitation, but to stimulate them to initiate new activities. Marler is tempted to strike an analogy with play. The vocal gyrations that birds go through during late sub-song and early plastic song are, in effect, vocal play, and play behaviour is one of the patterns of behaviour known to be influenced by the richness of the environment. This is particularly true in children. As the invented syllables are a very recent discovery, Marler has not yet adjusted to the novelty of the finding, but he is excited about any future research. So far they have discovered creativity in the song of a bird with one of the simplest of songs, the swamp sparrow. It is exciting to contemplate the degree of inventiveness to be discovered in the songs of mocking-birds or blackbirds. This is a promising area for future research.

If all song birds are learning songs at roughly the same time, why don't some of them learn the wrong ones? In the early days of research, the Cambridge chaffinches were presented with the song of the tree pipit *Anthus trivialis* and learned it. This was pursued further, by Peter Marler, with the white-crowned sparrows in California. They were given a natural choice of songs to learn.

In the coastal chaparral in which white-crowned sparrows live, another very common species is the song sparrow. The two species live closely together. As far as field research could show, white-crowned sparrows never learn song sparrow. Birds were brought into the laboratory. White-crowns were given a choice of recordings of white-crowned and song sparrows and unerringly imitated white-crown. If they were given song sparrow alone, the birds would reject it and revert to their innate song.

Marler was intrigued. Here was one of the most elaborate learning processes known in animals, and yet there appeared to be innate instructions or guidelines as to what the bird should be learning, an interesting interplay between nature and nurture. A bird cannot develop normal song without the opportunity to learn and yet it inherits some capacity to identify what it ought to be learning.

Marler and Susan Peters set out to see whether the swamp sparrows and song sparrows of New York had innate guidelines for song learning. Again, these two species live closely together in the same habitat, within earshot of each other. Instead of playing them tape recordings of natural songs, Marler began to synthesise songs using a computer.

Syllables from natural songs of the two species were edited out and recombined in different patterns, some swamp sparrow-like, others song sparrow-like. For every pattern devised two versions were created – one composed of mainly swamp sparrow song and the other of song sparrow. The research team gathered its battery of training songs, brought birds into the laboratory, and trained them between 20 and 60 days of age, the period when their readiness to learn is at its maximum. Then they waited to see what they would copy from the training songs. The results were interesting.

Take the swamp sparrows, for instance. Marler found that they were highly selective in what they would learn, and the selectivity was based, not on the pattern, but on the syllable. A swamp sparrow would learn any swamp sparrow syllable irrespective of the pattern in which it was presented.

This was the first robust demonstration of selective learning as an all-or-nothing phenomenon. Marler is now attempting to pin down the acoustic features on which a swamp sparrow makes its innate choice of what it should learn and what it should not learn. He has identified already the features that are *not* important. There is, for example, a slight difference in pitch between the two species but manipulating and swapping the frequencies did not make any difference.

Might there be something about the vocal tract that would make it physically difficult, if not impossible, for a swamp sparrow to sing the same syllables as a song sparrow? Marler devised a way of testing for this. It is possible to get a song sparrow to learn a song which has song sparrow patterning, but which is made up of swamp sparrow syllables. Marler took a swamp sparrow syllable and taught it to a song sparrow. He took the tape of the song-sparrow-sung-swamp-sparrow syllable, edited it out, and played it to a swamp sparrow. He could then see if the syllable had become unacceptable, having gone through the vocal tract of a song sparrow. The swamp sparrow learned the syllable.

To cut a long story short, Marler eventually concluded that an innate perceptual process must be at the heart of selective learning. Birds, from the age of a few weeks, have an innate perceptual filter in the brain which accepts or discards incoming auditory information. Other researchers are not so

Plate 1 John Burton, organiser of the BBC's Natural History Unit sound library, and Dave Tombs, senior audio assistant, deploy stereo recording apparatus in the field: (left) stereo parabolic reflector, (right) pair of gun microphones with wind shields.

Plate 2 Mark Sharp at the University of Hawaii signals the gesture for 'ball' to a bottlenose dolphin in dolphin–man language experiments.

Plate 3 Humpback whale 'breaching'.

Plate 4 Male blackbird on its songpost.

Plate 5 Barn owl with prey.

Plate 6 Male natterjack toad calling from the edge of the pond.

Plate 7 Female natterjack toad (right) approaching a calling male to initiate mating.

Plate 8 Merlin Tuttle in the rain forest of Barro Colorado observing mud-puddle frogs and fringe-lip bats with the night-vision-scope.

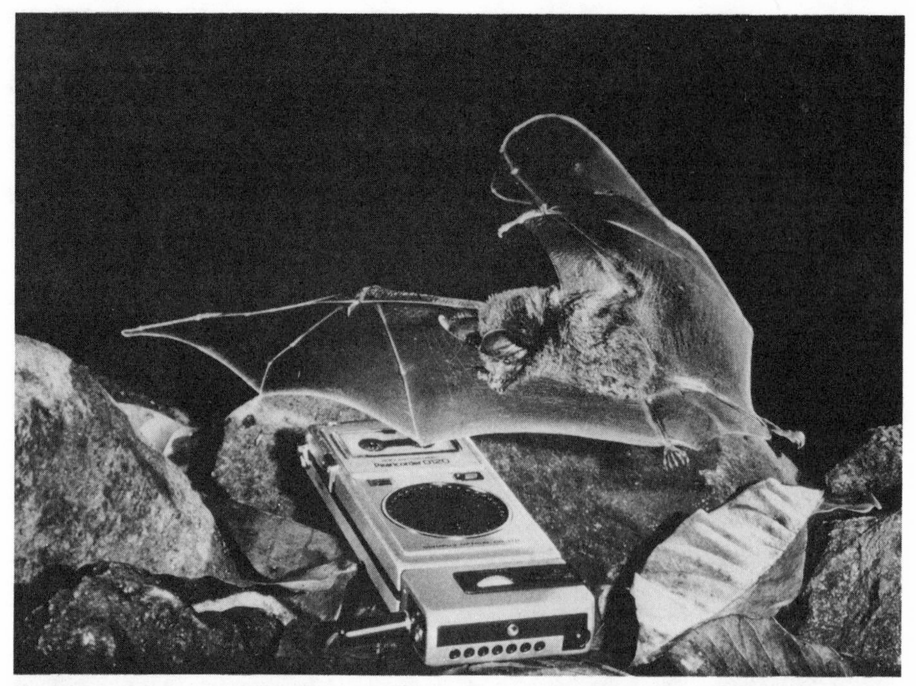

Plate 9 Fringe-lip bat responds to playback of mud-puddle frog calls and attacks the tape recorder.

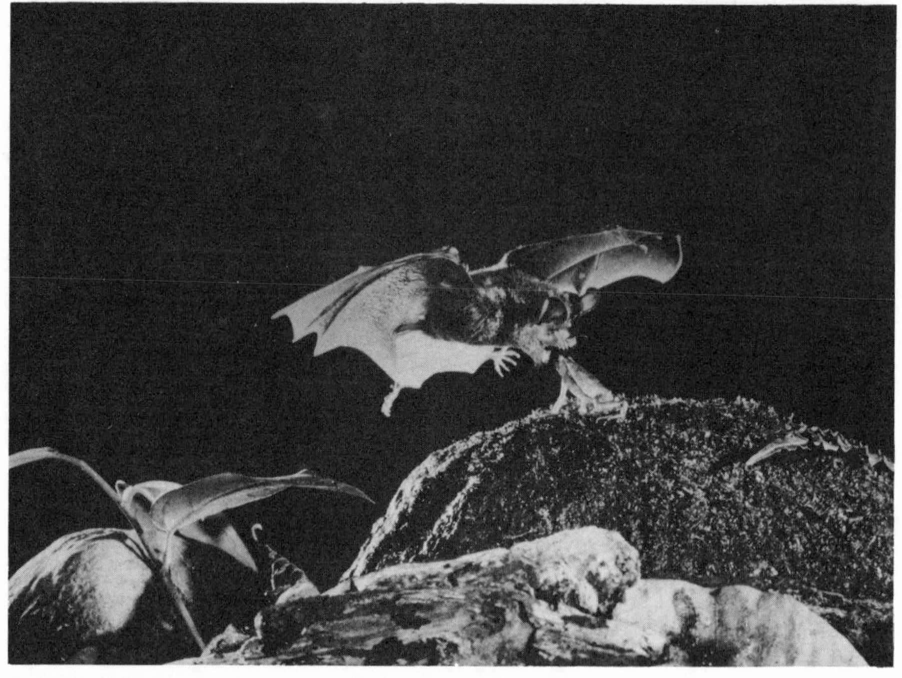

Plate 10 Fringe-lip bat takes mud-puddle frog from pond.

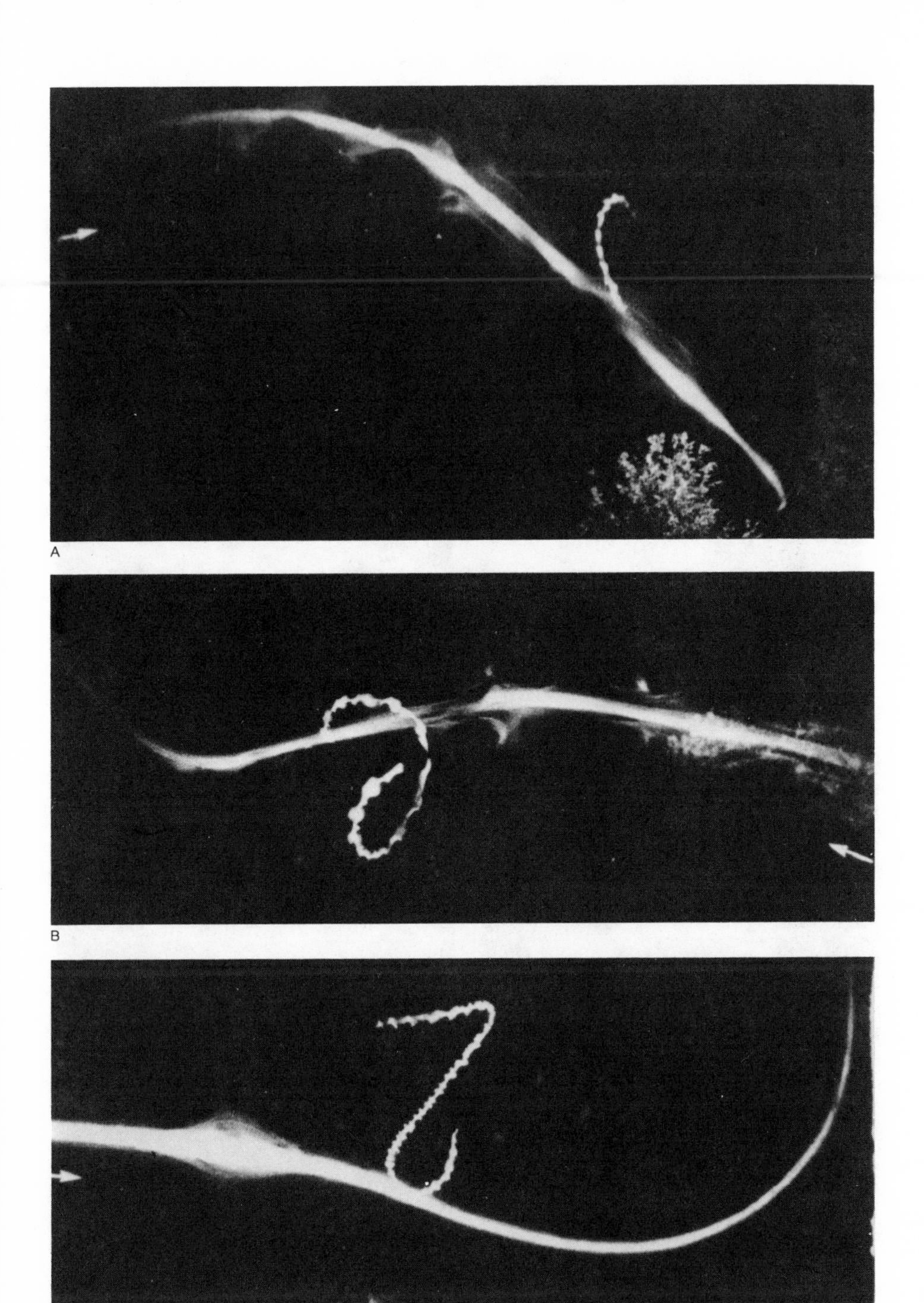

Plate 11 Streak pictures of bat–moth encounters: A Moth captured immediately after initiation of spiral dive; B Moth spiral is predicted by bat and moth is captured; C Moth evades bat with use of sudden secondary back-loop and the bat flies on.

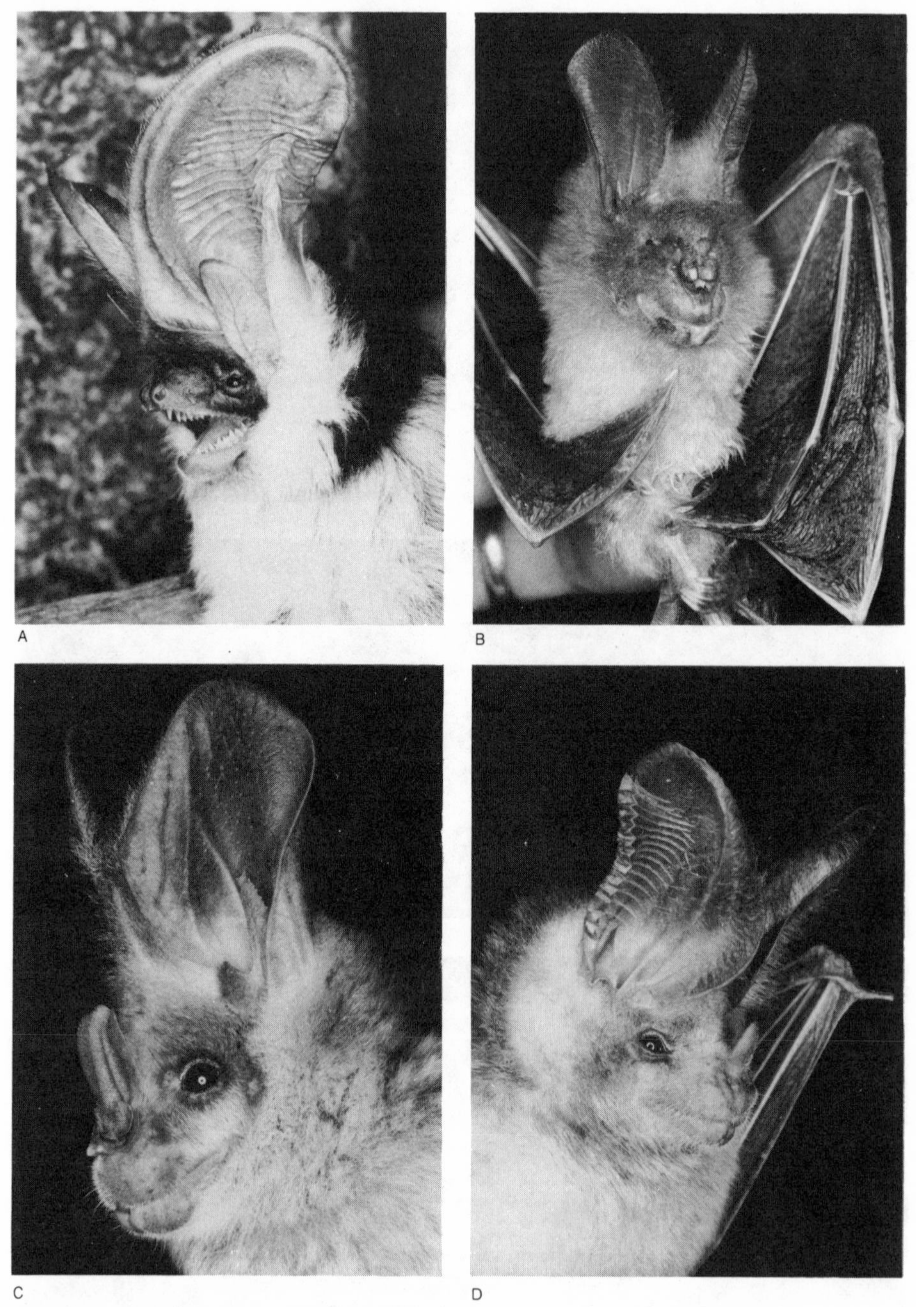

Plate 12 A Spotted bat *Euderma maculatum* of North America; B Large slit-faced bat *Nycteris grandis* of Africa; C Indian false vampire bat *Megaderma lyra*; D California leaf-nosed bat *Nacrotus californicus*.

Plate 13 Indian false vampire bat approaching food without using echolocation clicks, but craning neck to get better look at the prey.

Plate 14 Wolf spiders in the test arena. Female under the plastic bubble.

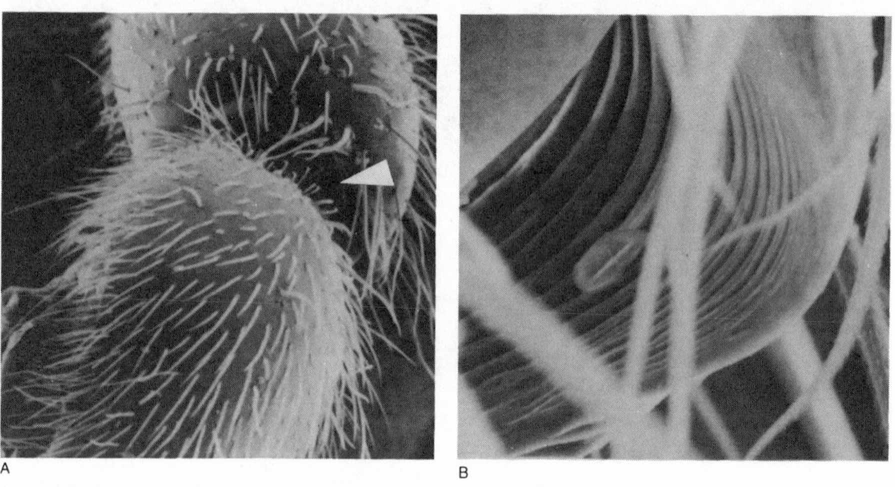

Plate 15 Scanning electron microscope photographs of male wolf spider palp, showing: A Location of stridulatory organ inside the joint of the tibia and tarsus showing the position of the file. The plectrum is hidden from view; B Enlargement of the file.

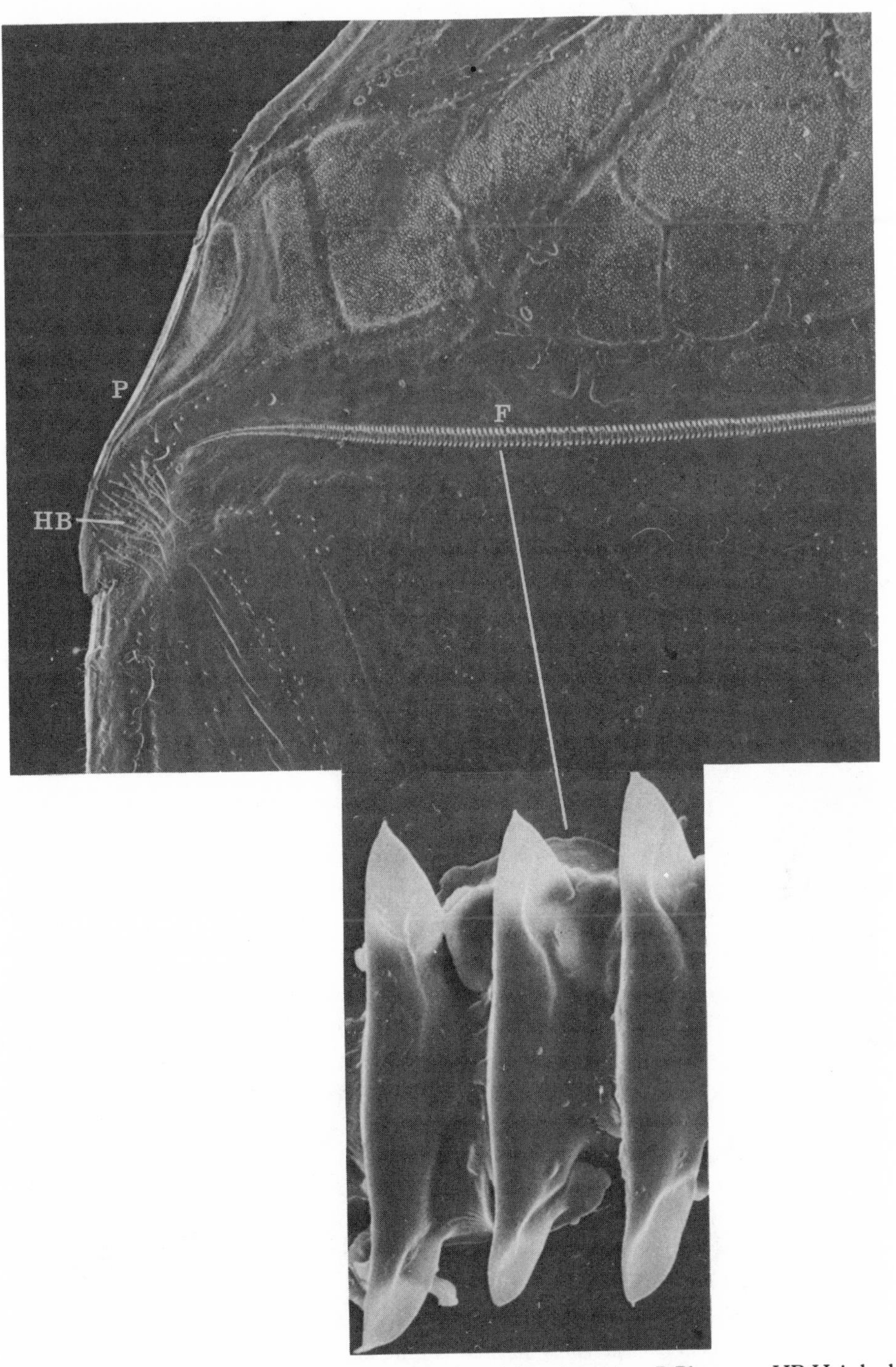

Plate 16 Teeth and plectrum of the house cricket *Acheta domesticus*: P Plectrum; HB Hair bed F File (enlarged below).

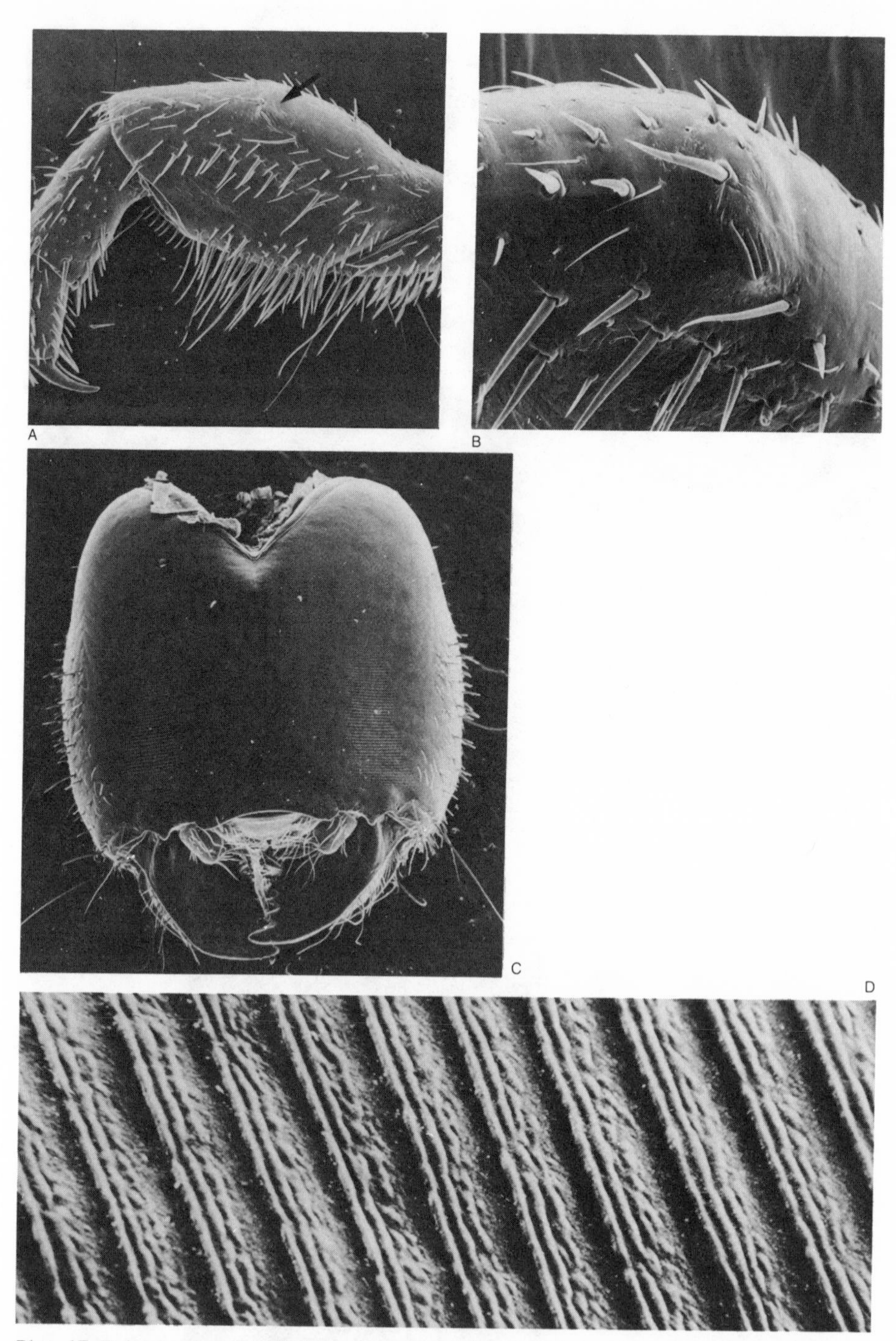

Plate 17 Enlargement of stridulatory apparatus of the caddis fly larva *Hydropsyche instabilia*: A Foreleg with plectrum; B Close-up of plectrum; C Underside of head with file; D Close-up of file ridges.

Plate 18 Red deer stag bellowing.

Plate 19 Red deer stag fighting.

Plate 20 Dog fox courting vixen.

Plate 21 Vervet monkey family group.

Plate 22 Dorothy Cheney at Amboseli.

Plate 23 An alert vervet monkey responds to an alarm call.

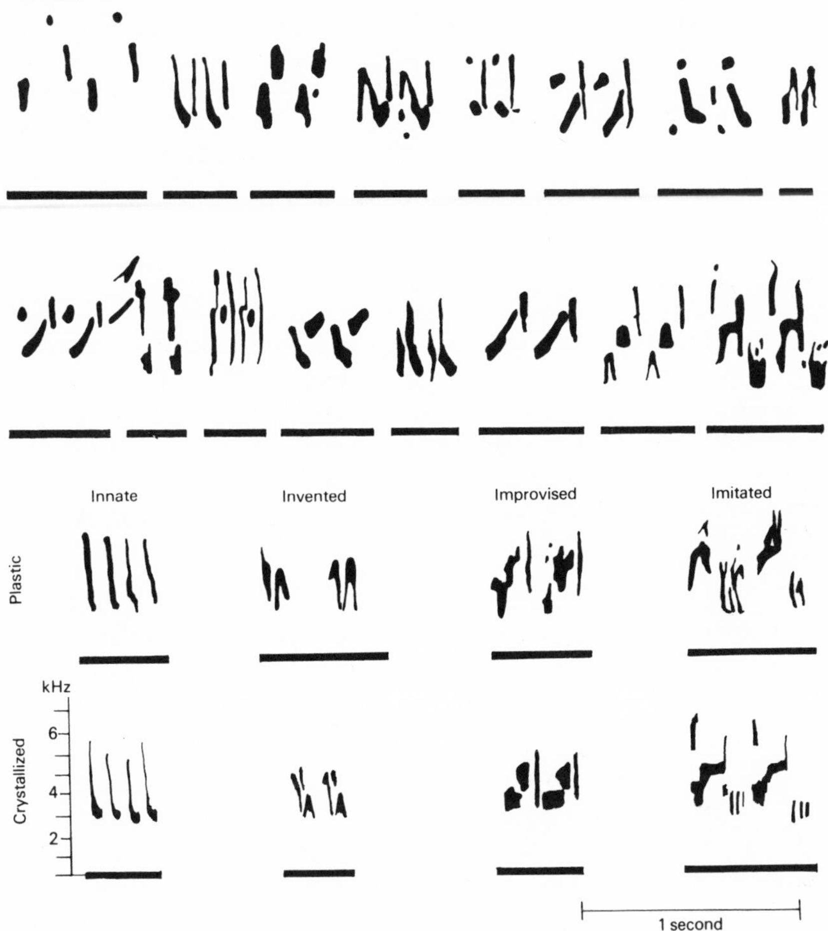

Fig. 22 Pairs of song syllables used for training swamp sparrows and pairs of syllables developed later in plastic and crystallised song, classified as inventions, improvisations and imitations.

convinced and suggest that the birds may *remember* everything but *reproduce* only the appropriate songs.

More recently, Luis Baptista, experimenting with white-crowned sparrows in California, has cast doubts on the universality of any innate filtering mechanism. He succeeded in teaching the birds all sorts of alien songs which, if a perceptual filter was working in their brains, they would have been unable to learn. The trick was to put the tutor and pupil in visual contact. Birds, it seems, are fussy about what they learn from recorded tapes and are more prepared to learn from individuals with which they can interact.

Marler, however, still believes that, given the multitude of environmental stimuli affecting young organisms, they must be provided with some innate instructions about how to sift the barrage of incoming information if their nervous systems are not to be reduced to chaos. To which stimuli should the creature attend and which ignore? At which point should it be ready to accept less favourable stimuli if the optimum circumstances do not arise? There must be innate instructions about timing, periods of life when an organism must go through a certain process of behavioural development to provide the skeleton on which later stages of development must depend.

Song Dialects

The consequence of birds having learned their songs is that individuals of the same species, living in different areas, will sing songs which are not identical. Just as local variations in human speech occur from place to place within a country sharing the same language, so too do some birds have small pockets of song variation – local dialects. Ornithologists in the San Francisco Bay area, for example, can place a white-crowned sparrow to within a couple of kilometres of its home range simply by hearing its song.

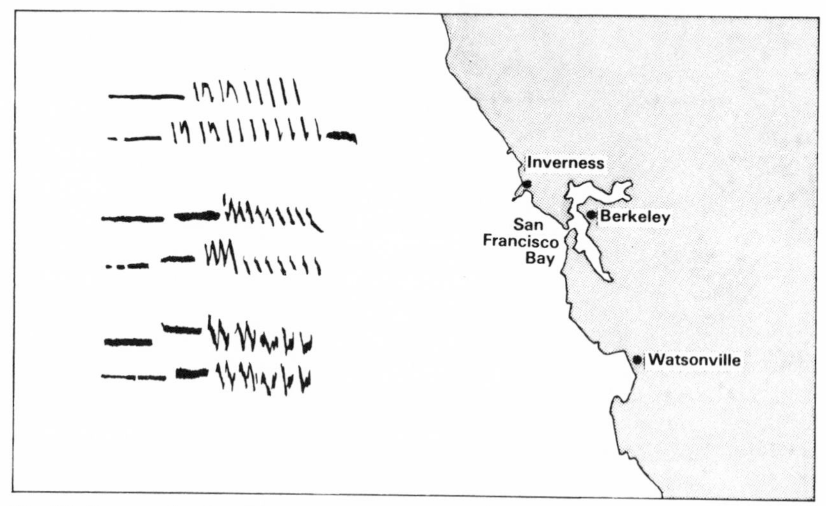

Fig. 23 Songs of male white crowned sparrows from three distinct dialect areas around San Francisco Bay.

When at Berkeley, Peter Marler, together with M. Tamura, analysed sonagraphically the songs of white-crowned sparrows and discovered considerable variation in the simple song. Basically, these birds have an alerting whistle, followed by a trill, with sometimes a buzz. The sparrows around the Berkeley campus sing a simple series of three whistles and end in a trill, whereas the Main District birds, to the north of San Francisco, have an opening whistle, a trill, and finish with a buzz. Individuals from Sunset Beach to the south, have a double whistle followed by the buzz and terminating in a trill. Throughout the Bay area there are quite distinct sparrow songs. In Europe, the same distinct dialect areas have been noted for redwings *Turdus iliacus* and corn buntings *Emberiza calandra*.

Marler and Tamura went further, and tested white-crowned sparrows for evidence that dialects were, indeed, a result of cultural transmission during song learning. Hatchlings that had not heard adult song and young birds that had were brought into the laboratory and then reared in social isolation. The hatchlings deprived of adult example sang songs quite unlike their home dialect, whereas the birds that had heard the songs of their adult neighbours eventually sang good copies of the local dialect. Tutoring studies later showed that sparrows would learn, during the critical learning period, the home dialect or even the dialects from other areas. This was evidence that dialects are learned; but what of their function?

Could it be that dialects are simply accidental occurrences as a result of learning and have no biological function? Marler and Tamura have speculated that female sparrows might use the song dialect for selecting a mate. The dialect might represent a local population indulging in some degree of inbreeding, and therefore, in a sense, could be considered as some form of incipient speciation, with the advantage that, as it is a learned behaviour, it could be reversed during the course of a generation. Unfortunately, work by Luis Baptista has confused speculations of this sort in that female white-crowned sparrows given testosterone to encourage them to sing, surprisingly sang with a different dialect from the one in which they were brought up!

Myron Baker, of Colorado State University at Fort Collins, took another line. He wanted to see if any genetic variations coincide with a boundary between two dialects. Using biochemical analyses to detect polymorphisms in the blood, Baker examined the genetic make-up of birds across a dialect boundary. He discovered genetic discontinuities which, he argued, are a consequence of males and females settling in an area where their home dialect is sung. He pursued this further by banding birds which had hatched along the transect and followed their patterns of dispersal. He found that birds in the heart of a dialect area disperse more or less radially to establish territories and breed, whereas those at the boundaries dispersed asymetrically, veering away from the boundary into the focus of the home dialect. Baker had revealed that the dialects of his study group of a sub-species of white-crowned sparrow form a kind of genetic mosaic. They are all members of the same species yet there is enough differentiation for each dialect area to constitute a deme or selective unit. This would also mean that birds living close to each other tend to be kin, a situation which favours some of the elaborations of social behaviour, such as reciprocal altruism, emerging.

In the light of these kinds of observations it has been suggested that dialects *do* have a function. Fernando Nottebohm, now at Rockefeller University, New

York, proposed a function for dialect when studying the rufous collared sparrow *Zonotrichia capensis* on the Pampas of Argentina. He suggests that the female of a species such as the rufous collared sparrow which has a wide distribution in a variety of climates or habitats from cold tundra to hot desert, would benefit by mating with her own dialect group. A bird adapted to a desert would best pair with another desert bird of the same species. Mating with a bird from a polar area, for instance, would dilute the gene pool, possibly introducing inappropriate characteristics, and might make the offspring less likely to survive the rigours of the desert. The rufous collared sparrow bears this out. Across large tracts of the Pampas, with uniform plains vegetation, the songs of the sparrows are very similar. The birds in the hills, where the climate and vegetation change, have quite different dialects.

But, although the work of Marler, Baker and Nottebohm has suggested a function for dialects in white-crowned sparrows and rufous collared sparrows, Peter Slater, of the University of Sussex, is reluctant to extend these findings to song variations in many other species. Slater feels that the tremendous amount of emphasis put on dialects and geographical variation in bird song has been overdone. His interest is in chaffinches, and he feels that variations in their songs serve no function whatsoever. In the case of the chaffinch, dialect-like differences really are an accidental by-product of song learning.

A young bird, within the first year of life, hears other chaffinches singing. He listens and learns, memorising the exact details heard. The following spring, when his testosterone levels begin to rise, he starts to sing himself, matching his own output to his memory of what he has previously heard. Slater has demonstrated that hand-reared chaffinches, trained in the first two months of their life with song types that they did not hear subsequently, produced near perfect copies six or eight months later. The problem about learning as a strategy of song development, suggests Slater, is that animals make mistakes, and mistakes in learned patterns are much more common than mutations in genetic transmission. As a result song is almost certain to change slowly.

Some individuals copy accurately, others not so accurately. Some song types may be copied less than others and become extinct; others are mis-copied, so creating new songs. If populations are distinct from each other, then birds will learn primarily within their own population, and as a result the song and their mistakes will be different from those of a population some distance away.

In the wind-swept Orkney Islands, off the north of the Scottish mainland, Slater has been studying a group of chaffinches and looking for local song variations. There are few trees on the Orkneys, consequently few chaffinches, so Slater has been able to record a very high proportion of all the birds in the area. He found that, not only was there more than one song type for the whole of Orkney and several song types for each major wood, but there were also individual variations.

Each chaffinch has between one and five song types. There are song types shared by many birds and other song types singular to a particular bird. In one wood, for example, with 15 male chaffinches, Slater identified ten individuals singing song-type B – the phrases within the song were almost identical with few variations. But another bird might have a type very similar to B, but not quite the same. Slater suggests it had copied song type B but had made a mistake.

From the 40 male birds in the study area, Slater was able to recognise 17 different song types. If every bird had its own songs the number would have been nearer a hundred, but this was not the case. There was obviously a degree of overlap of the song types. The grouping, though, is not strictly into dialects. If it were, one strand of trees would have a particular collection of song types, while the next wood would have a quite distinct set. This is not what happens. Slater found that songs tend to have foci. When the song is common about half of the population are singing it. As you get further from the song focus fewer birds are heard singing that particular song and another song type takes over. Travel a little further and you reach the focus of the new song. In chaffinches there are no distinct dialect boundaries. Rather, there is a patchwork of overlapping song types where some song foci interact with others, and where some song types are common to all areas. The song simply carries geographically from one area to another. Peter Slater feels that the word 'dialect' should be reserved for those species where there are areas in which a particular song type is sung, marked by border lines beyond which the song is not sung and another song is sung instead.

Vocal Copying

One way in which a song bird can acquire an elaborate song is to borrow elements from the songs of others – vocal copying. After all, song birds mostly learn their species song originally from neighbours. This kind of copying from males of the same species is the first of four categories of vocal copying in song birds described by David Dobkin, of the University of California at Berkeley. He calls this kind of infraspecific copying 'vocal imitation'. It is the mechanism of dialect systems. Dobkin thought that the literature was littered with a host of interchangeable names that described any form of vocal copying – vocal mimicry, vocal imitation or vocal convergence – and was determined to clear things up in order that scientists would be talking about the same piece of behaviour.

Dobkin came to this task when he became interested in the problem of two relatively closely related species living in the same area. If they are breeding within earshot of one another, using songs to keep rivals of their own species at bay, and to attract females of the right species, what are the implications for the kinds of signals being transmitted? The birds are potentially close competitors, particularly if they are morphologically similar. They might be feeding at the same time, on the same kinds of foods and have similar nesting requirements. Species A needs not only to keep out other species A males, but also to chase away species B males. This also raises problems of unambiguous recognition of signals on the part of each species. If species A vocalisations are similar to those of species B, then species A males are likely to attract females of species B, resulting in no offspring, or at the last, infertile hybrids. This is not advantageous, clearly, for the individual, so there should be some selection for species specificity. But, at the same time, if a bird is in a situation where it may increase its own nesting success by excluding potential competitors of another species, then there is opposite selection pressure for some similarity of signal. This second category of vocal copying Dobkin has called 'vocal convergence', and it occurs when ecological competitors or close relatives have similar songs or calls.

Dobkin termed his third category 'vocal appropriation'. This happens when an individual copies sounds made by members of another unrelated species. It is often accidental, as when an individual incorporates into its song elements of another species' song simply because it lives in the same acoustic environment. Vocal appropriation may occur as a result of an impoverished acoustic environment. If a species is nesting in an area at a very low density, and there are not many role models around from which young birds can learn, then a youngster may pick up song elements of a close neighbour of a different species.

In the literature vocal appropriation has often been described as 'vocal mimicry', but Dobkin is fast to point out that mimicry involves deception, and deception is probably not intended in this form of copying; it is simply a way of enriching the song repertoire.

Vocal mimicry is, in fact, Dobkin's fourth category of vocal copying, but he questions its existence in animal vocal communication. There is, he believes, only one documented study in which vocal mimicry takes place. This is the case of the violaceous euphonia *Euphonia violacea*, a member of the tanager family. If a predator approaches the nest, the resident pair give the alarm and mobbing calls of other species which are attracted to the nest and mob the predator.

Some birds copy bizarre sounds. Blackbirds have been heard to copy 'whistling' telephones and pedestrian-crossing bleepers. Why they should do this is a mystery. On an old Ludwig Koch recording, made in 1910, a celebrated Prussian blackbird even imitated the distinctive sound of the Kaiser's motor klaxon. The superb lyrebird *Menura novae-hollandiae* of Australia regularly incorporates farm machinery noises into its courtship display. An even more puzzling form of elaboration to a bird call occurs when captive hill mynahs *Gracula religiosa* accurately copy human speech. In the wild, copying is basically confined to species calls or the calls of neighbours, although there are many exceptions. Mynahs often incorporate the staccato sounds of local tree frogs into their complex calls.

Duetting

In some species, for example tropical boubou shrikes *Laniarius ferrugineus*, an elaborate song is built up from the contributions of two birds. Instead of the male singing alone, the female joins in too. The coordination and integration of the two songs is so well achieved that it often sounds like the song of just one bird. How the coordination is achieved we do not know.

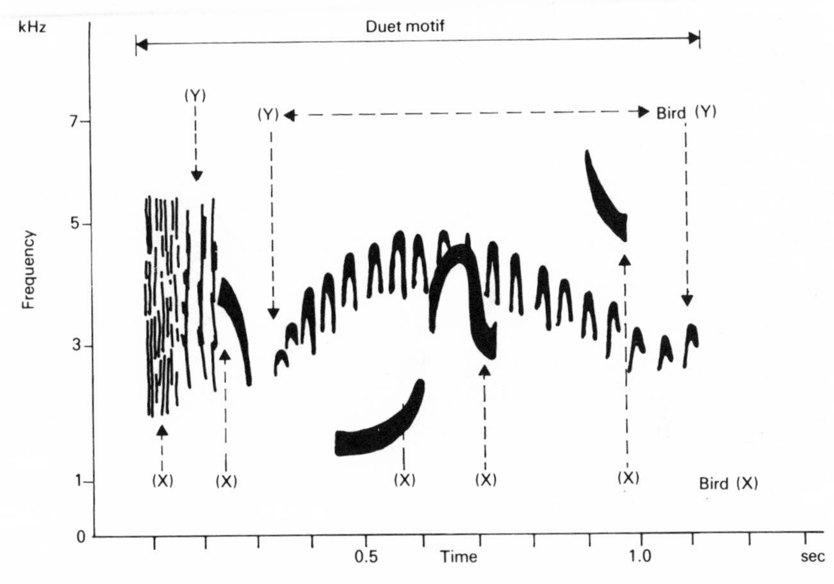

Fig. 24 Duet motif of a pair of *Cisticola hunteri* from Kenya. Bird (x) is thought to be female and (y) male.

Why should birds want to duet in this fashion? A clue may lie in the habitat of most duetting birds, for they live almost exclusively in tropical forests or other places with dense vegetation. They also tend to be monogamous – pairing for life – and to retain territories throughout the year, for many years. Duetting, perhaps, maintains and reinforces the pair bond. In dense vegetation with birds constantly out of sight of each other, a sound reinforcer would be important.

Duetting birds are often seen singing at other duetting birds across a territorial boundary. Duetting, then, like solo song, may have the dual function of establishing and reinforcing the relationship between a pair and of territorial proclamation.

Song Production and Control

In order to make their beautiful songs, birds employ some clever anatomical tricks. Basically, air from the lungs is forced over vibrating membranes – the syrinx – much as in our own larynx, but there the similarity ends. The bird syrinx is a little more complicated.

Unlike the human larynx, which is at the top of the trachea or windpipe, the bird syrinx is at the bottom, where the two bronchi meet coming from the lungs. The syrinx sits astride the junction. In many birds there are separate vibrating membranes or chords on the two arms of the syrinx, and as a consequence they have the remarkable ability to produce two sounds at once. A bird can sing two quite distinct tunes at the same time.

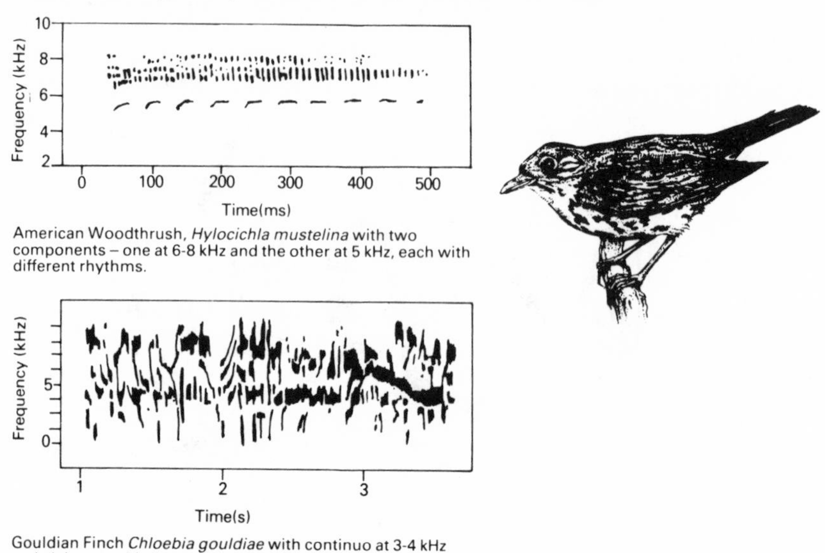

American Woodthrush, *Hylocichla mustelina* with two components – one at 6-8 kHz and the other at 5 kHz, each with different rhythms.

Gouldian Finch *Chloebia gouldiae* with continuo at 3-4 kHz and elaborate chirping.

Fig. 25 Complete songs of birds showing contributions from left and right half of syrinx.

Another difference from human speech production is that much of the structure of speech results not from variations in the voicebox but from movements of the tongue, mouth and nose, which change the resonant properties of the various cavities. Bird song is not produced that way. Most birds are not dependent on the resonance of the vocal tract. Instead, the entire system is driven actively by the membranes of the syrinx. American thrushes are fine examples of birds that can sing two songs simultaneously, but with a harmonious relationship between each rendition, producing some quite extraordinary tonal qualities which, some researchers feel, are not to be matched anywhere else in bird song.

A mature song bird, then, has two sets of contributions to its song – one coming from the left side of the syrinx and the other from the right – and they are precisely coordinated. One of the problems a young bird must overcome is how to achieve that coordination. Fernando Nottebohm, of Rockefeller University, New York, studied the nestling calls or the fledgling calls produced by a nestling chaffinch, and could distinguish between the contributions from the right and left side. He found that, in the young bird, the relationship between the two sides of the syrinx is not controlled. The coordination varies, giving the song a harsh quality. There is evidence that the fledgling call of the chaffinch provides practice in syrinx coordination, which is a necessary prelude to the process of singing and of song learning.

Nottebohm followed the work through with domestic canaries and showed that there is a parallel between bird song production and human speech production in that one side of the syrinx tends to make a dominant contribution. In the case of the canary, it is the left side. And the dominance is not only evident in the syrinx. Dominance of the left side can be traced up into the brain – an analogy with the phenomenon of hemispheric dominance in the control of human speech. This is the only animal analogue for this phenomenon studied so far and it has been the thrust of Fernando Nottebohm's research for the past ten years.

Birds, we have seen, learn to sing by reference to auditory information. There is a set of built-in rules that specify the kinds of auditory information to be accepted, and a repertoire is developed by reference to these auditory expectations. Normally the song manifests itself in imitations of other adult models. How, then, is the brain handling this? In principle, there should be an efferent motor system and an auditory feedback, which at some point must integrate, so that when the song is matched to the auditory template, and learning completed, the pattern may be held and stored in long-term memory. Nottebohm wanted to find out which part of the brain carried out these tasks. He started by looking at the motor part of the auditory loop as that was more accessible. Attention was focused on the syrinx.

Nottebohm and his colleagues Christiana Leonard and Tegner Stokes, discovered three regions of nerve cells in the brain which seem to be devoted, almost exclusively, to the control of the syringeal muscles, and which he assumes also control aspects of song learning related to the performance of the syrinx. These song nuclei are simply clusters of cells which are quite distinct, with discrete boundaries, and which can be easily identified with specific staining techniques. Two are located in the forebrain, the telencephalon, and one in the hypoglossus. The hyper-striatum ventrale pars caudale known as the HVc at the top of the forebrain is linked to the nucleus robustus archistriatalis (RA), which in turn tells the neurones that enervate the muscles (hypoglossal motor neurones) to organise and activate the muscles of the syrinx.

Nottebohm carried out some simple surgery on the brains of canaries. If he lesioned the top nucleus, HVc, the birds would continue to sing, but they did so in a virtually silent fashion. The dynamics of song were expressed but not the sound itself. The throat quivered, the bill held slightly open, the body feathers were sleeked and the bird turned around and directed its behaviour to another individual in another cage. All that was heard was a faint clicking sound, which corresponded to the temporal pattern of the song, but there was no frequency modulation, none of the sophisticated patterning characteristic of canary song. It was, Nottebohm recalls, as if the motivation for singing was there, much of the motor control for song was present, but the syrinx did not function. Nottebohm had demonstrated that the three-nuclei auditory pathway is primarily concerned with song and call production.

In canaries, it is mainly the left half of the syrinx that is involved in singing. The right half does very little; indeed in some species it does not play any role in sound production. The nerves running from the brain to the syrinx and back are ipsilateral, that is, the song nuclei in the left side of the brain control the left side of the syrinx. This is unlike other motor pathways, which are crossed, so that one brain hemisphere controls the behaviour of the other half of the body. In humans, for example, the left hemisphere controls what the right hand is doing. In canaries the left hemisphere is dominant for song production.

If song nuclei on the right side of the canary brain are lesioned singing is barely affected: some syllables tend to be unstable and a few disappear, but the phrase structure and repetition of syllables typical of canary song persists. There was, however, a surprise in store for Nottebohm and his colleagues. Although there is marked dominance of the auditory system by the left side, if a canary is left for a few months, gradually the right side will take over. So, even though the bird has two parallel pathways for song control, and even though the left side is dominant, it has a right side which is underused, but can do the job just as well.

Left hemisphere dominance seems to be the case also in chaffinches, white-crowned sparrows, Java sparrows *Padda oryzivora* and white-throated sparrows *Zonotrichia albicollis.* But not all birds are left-handers – zebra finches appear to be right-handers. Within a species there is a strong tendency to consistency.

If brain nuclei control a bird's singing abilities, an activity taking place mainly in spring and early summer, it might be expected that the nuclei change with the seasons. Nottebohm looked to see if canary song nuclei alter during the course of the year. He found that in the spring, when canaries sing a lot, the nuclei are very large. In the autumn, five months later, at the end of the moult, the nuclei are half as large. The observations are relatively new and go no further, but Nottebohm speculates that the reduction in size is linked to a reduction in the amount of time devoted to the control of singing behaviour. He suggests that the network is partially dismantled. Microanatomy by Timothy DeVoogd has shown that the dendrites, the fine processes on brain cells, are absorbed. This is in line with the observation that a canary learns a completely new song repertoire each year. Maybe, proposes Nottebohm, the bird is rejuvenating its brain pathways for song control, so that when it starts a new episode of growth, the network is ready for song learning once more. The researchers are now looking for the exact point during the five months between spring and autumn when the changes take place. The hunch is that it could be closely associated with the moult, when all kinds of physiological changes are taking place. Nottebohm is tempted to think of the moult as being a time when extra network space is thrown away to make room for the new wave of learning.

It might also be expected that song nuclei would be of different sizes in males and females. Arthur Arnold, working with Nottebohm, found that the song nuclei of canaries and zebra finches, particularly those in the forebrain, were several times larger in males than in females. At the time, the orthodox view was that the brains of males and females in any vertebrate species were virtually identical. Only small differences, requiring microdissection and electron microscopy to detect, were thought to occur. In the canary the sexual dimorphism is so marked that a slice of brain tissue, held up to the light, will show the difference clearly if the nuclei are stained a dark colour. This seemed to make sense to Arnold and Nottebohm. The male canary spends a good deal of time learning to sing and then sings in the spring. The female does not sing at all. It looked like good brain economics that if the female does not indulge in a particular piece of behaviour then she should not devote much brain space to its control.

The researchers went one step further. Male canaries have great variability in their songs. Some have song repertoires three times larger than others. Arnold and Nottebohm looked to see if repertoire size was correlated with nuclei size. They found that birds learning larger song repertoires have larger forebrain song nuclei in the brain.

Nottebohm calls this the 'library principle'. He suggests that learning is like the storing of books in a library. If you want to store a lot of books you need a lot of shelf space. You may, on the other hand, have a lot of shelf space and few books. Size, he feels, acts in a permissive manner. If a lot of skill for a particular problem is developed then it helps to have a lot of brain space devoted to the task. If the brain space at the outset is small, then the chances of excelling are small.

Interestingly, female canaries can be induced to sing with hormone treatment. Three weeks after injection of the male hormone, testosterone, females will begin to come into song, but the complexity of the song is that which would be expected from the small size of their song nuclei. Instead of 30–35 different syllables, they only sing five to ten different syllable types. One other interesting observation is that after testosterone has been given, the dendrites of the brain cells in the song nuclei proliferate, presumably adding to the network space available in the brain to control the new behaviour. Nottebohm postulates that, in spring, the male song nuclei also increase the number of dendrites with the secretion of testosterone. He also speculates that there may be an increase in the number of actual brain cells, a revolutionary proposal. An increase in the number of cells has been found but further work must be carried out to ascertain whether they are neurones or other types of cells.

The medical implications of this work are also being considered although the parallel is not close enough for immediate application. It is unlikely that in humans a reproductive hormone, like testosterone, would be found to be involved in regulating neuroplasticity. Humans are essentially not seasonal creatures in the same way as birds. There are likely to be, however, hormonal factors involved in processes of maturation as the young infant develops, stages that correlate with babbling, and it might be possible to induce the same local hormone condition in the adult brain of someone suffering a stroke. A cannula packed with hormone crystals, for instance, placed on a small part of the brain, might bring about hormonal conditions that mimic those of earlier development, leading to rejuvenation of the damaged site. This, Nottebohm emphasises, is mild speculation but it is, nevertheless, a distant possibility.

7

MOTHERS AND BABIES

Crocodiles

The crocodile has survived, little changed, for 200 million years. It is a living fossil, one of the few remaining members of the group of animals that included the great dinosaurs. The crocodile shows, to a remarkable degree, parental care of both nests and young.

After courtship, pairs of Nile crocodiles *Crocodylus niloticus* stay together. The male patrols the nest site, and is often seen cruising off the breeding beach. The female lays up to 90 eggs in a pit in the sand, by the side of a lake or river. The nest pit is covered and guarded by the female for about four months, until the young are ready to emerge. During the entire period she does not eat. The eggshell, during the incubation period, is gradually weakened as nest bacteria progressively dissolve the calcium in the shell, increasing the porosity to allow the youngster to breathe, and decreasing the strength so that it may break out. At hatching time the nest is activated by some hidden signal. From below the sand can be heard a chorus of 'chirping' calls of the young crocodiles. They call for their mother to come and release them from the nest. When the youngsters detect the vibration of the mother's footsteps they begin to call. If she stops, they stop. When she walks forward they start to call again. The mother digs down, scrapes the sand away and, one by one, as the eggs are exposed, a little snout with its eggtooth cracks open each egg and the youngster appears.

At the St Lucia and Ndumu Game Reserves, in Zululand, Natal, South Africa, Anthony Pooley has been watching the nest opening behaviour of Nile crocodiles. He has shown that the stimulus to the female provided by the 'chirping' young crocodiles, still in the nest, is very strong indeed.

As part of the Natal Parks Board's crocodile rearing and research programme, Pooley and his colleagues were studying captive crocodiles. One female laid eggs which were infertile and did not hatch. She was chosen for playback experiments. Recorded tapes of hatching crocodiles were made. When played back on loudspeakers placed just outside the enclosure, the female rushed from the pool, headed straight for the sound and started to scrape vigorously. If the tape recorder was switched off she would stop; if switched on again, she would try to release the babies behind the fence.

It is thought that hatching usually occurs at night, when the baby crocodiles are less likely to fall prey to aerial hunters, such as hawks and eagles. The female identifies a suitable nursery area next to the lakeshore or river bank and carries the hatchlings gently to the water's edge.

Many newspaper articles in the past have shown crocodiles with babies inside the mouth cavity. Cannibals! they were labelled; but nothing could be further from the truth, for crocodiles are admirable and diligent parents.

The female, attracted by the call, releases the youngsters, but only a few at a time. The young crocodiles 'chirp' continuously and flick their small tails to attract the mother's attention. She picks them up one at a time and allows them to sit in a special pouch in the floor of the mouth. The calls become less frantic as each youngster is safely aboard. The female carries them to the water, opens her mouth, and they swim out, again calling excitedly to contact their brothers and sisters in order to keep the group together. The female then goes back to the nest site and digs out another batch which, by this time, may be giving a more strident, double syllable distress call. The sound can be heard 30 metres away, so it can be heard very clearly by the female at the water's edge. Each time she hurries back. As each group is introduced to the water the others give a greeting response, a soft cricket-like contact call. These repeated trips ensure that no youngsters are left exposed, in the open, to predators. The eggshells and membranes are eaten by the parents to avoid attracting predators. If a fish eagle, for example, does fly over, the mother crocodile's body quivers and, as one, the baby crocodiles dive below the water. The youngsters stay with their mother for several weeks during which they learn to feed efficiently, and to develop their own ways of escape when danger threatens.

Pre-hatching and Pre-birth

For many animals, particularly birds and mammals, communication by sound starts some time before the youngster is hatched or born. Young quail chicks, for example, synchronise their hatching by calling.

During incubating and brooding, female mallard ducks *Anas platyrhynchos* make a species-typical maternal call to which the unhatched ducklings respond. The mother does not start making the call until the embryo duck's head projects into the airspace of the egg, at about 17–19 days after laying. The hatchlings, then, are aware of their species call some time before they emerge and respond immediately to the maternal call when called out of the nest.

At the University of Maryland, Slobodan Petrovich and Eckard Hess revealed that ducklings exposed to the mother's clucking before they had hatched would respond to decoys emitting the species call, and would imprint on the decoy; but they would not react to silent decoys. David Miller and Gilbert Gottlieb of the North Carolina Division of Mental Health, at Dorothea Dix Hospital, Raleigh, discovered that repetition rate and frequency modulation are the salient features, although the calls used before and after hatching are slightly different, in that there are more notes per burst and more harmonics in the call given after hatching.

Working with Timothy Johnston, Gottlieb found that the mother's cluck could be more important to the young duckling than her visual appearance. In experiments, ducklings were visually imprinted on a stuffed duck model soon after they had hatched. They were then presented with a red-and-white striped box and the stuffed model, both of which had loudspeakers hidden inside. If the speaker inside the box played the mother's 'assembly' call the ducklings would rush towards it, totally ignoring the more visually attractive stuffed bird.

A. E. Storey and L. J. Shapiro, of the University of Manitoba, carried out similar experiments, but replaced the stuffed model with a live, silent female duck. The ducklings still went to the sound source. If the researchers presented the ducklings with an additional model – a live female plus maternal call – then the ducklings preferred this combined audio-visual stimulus to any others.

As far as the ducklings themselves are concerned, they emit two quite distinct calls – the 'contentment' call and the 'distress' call – which play an important role in keeping the family group together. The 'contentment' call is first given by the embryo in the egg. It has short note durations, fast repetition rate and is low pitched. Later, it is given in the presence of other ducklings and when the mother's call is heard. The 'distress' call is only heard after hatching. It has longer note durations, slower repetition rate, and is higher pitched. It is uttered when a duckling has become isolated from the group.

Parent-offspring recognition is not instant upon the hatching of the chick, but is dependent upon the lifestyle of the species. Experiments in which young of various ages are swapped between nests have shown that in colonial birds, such as gulls, brood exchanges can be tolerated until the chicks are about to move out of the nest. In the herring gull *Larus argentatus*, this parent-offspring recognition takes place about five days after hatching, whereas in the kittiwake *Rissa tridactyla*, recognition does not become effective until five weeks.

Pre-parent recognition begging behaviour was found to be three times longer and more pronounced in island fledglings of the varied tit *Parus varius*, than in mainland birds. The work was carried out by Hiroyoshi Higuchi of Tokyo University, and Hiroshi Momose of The International Christian University, Tokyo, who feel that the exaggerated behaviour and extended parental care have evolved as a result of the low availability of food and high density of birds on islands.

Michael and Inger Beecher and Shai Hahn, of the University of Washington at Seattle, looked at parent-offspring recognition in bank swallows (sand martins) *Riparia riparia.* Exchanges of like-aged chicks showed that in this species transfers could be made without upsetting the parents until 15 days after hatching. They would be rejected at 16 or 17 days. At 15 days an immature begging call changes to a 'signature' call, each individual having its own distinct call. In another study, the Seattle researchers found that parent birds would only approach a loudspeaker playing the 'signature' call of their own chicks.

Studying the protective behaviour of parent ring-billed gulls *Larus delawarensis*, Michael Conover, F. Dudley Klopfer and Don Miller of Washington State University at Pullman, showed that straying chicks give audible cues to their parents. The chicks would always wander off in the same direction so a parent gull would know in which direction to start looking for offspring. On finding a chick the parent would only respond if it was its own chick emitting its own particular call. Ring-billed gulls nest in dense island colonies. They are highly territorial. Neighbouring adults might kill a straying chick so it is important that contact between parent and offspring promotes protection for the chick. With unhatched eggs or young in the nest, the adult will give the 'long' or 'mew' calls. Chicks respond to the 'mew' call by returning to the nest site. 'Mew' calls are given during the feeding of the young. Roger Evans, of the University of Manitoba, believes that 'food' is the underlying motivation which encourages the chicks to respond to the parents' call. Playbacks have shown that laughing gull chicks *Larus altricilla* and razorbills *Alca tarda* can discriminate the voices of their parents from those of other adult birds.

Prehatching sound experiences play some part in the survival of chicks. In an experiment with gulls eggs, one clutch was exposed to maternal feeding calls, while another clutch received no sound stimulus at all. After hatching it was found that those chicks which had heard the mother's feeding call pecked at her bill for food. The other chicks did not. Margaret Vince, of the Institute of Animal Physiology, Cambridge, set out to discover whether pre-natal sounds were important to young mammals, too.

In separate rooms, two groups of pregnant guinea pigs *Cavia porcellus* were exposed to different pre-natal sound experiences. One group heard the feeding 'clucking' of a bantam hen, the other heard no sound at all. The adults hearing the hen would be startled, the animal would freeze and the heart rate would slow down. Repetition of the stimulus resulted in gradual reduction of the response until it was ignored. After the birth of the pups, the two groups of youngsters were presented with the bantam hen sounds once again and the heart rates monitored. The response of those youngsters born to parents that had been exposed to the sound was less pronounced than that of the control group. The control group's heart rates decelerated in the way they would if the youngsters were confronted with a strange animal. The experiment had shown that a youngster might become habituated to much of the normal sound environment of the mother.

Sally Armitage and B. A. Baldwin, together with Margaret Vince, then tuned in to the foetal environment of sheep. Hydrophones implanted inside the intact womb recorded the kinds of sounds that the foetal lamb might hear. The mother's eating, drinking, rumination, breathing and muscular movements were discernible, but the heart and blood systems, usually thought to contribute substantially to the background noise, were in fact not perceptible. Hydrophones attached to the necks of foetuses in two pregnant ewes revealed that sounds from the outside world can be heard clearly. Just as mother birds have auditory exchanges or 'conversations' with the embryo in the egg, so too, it is thought that mammals have the same capability. In the sheep the auditory system is functioning before birth, after day 100 of gestation. The experiments led the researchers to conclude 'that the auditory experience of the foetal mammal may be considerably more extensive, more varied, and, as in birds, possibly of greater post-natal significance than has been believed'.

Psychologists Anthony Delasper and William Fiter of the University of North Carolina at Greensboro have found, for instance, that human infants only three days old can distinguish their mother's voice. In the experiment, babies were invited to suck at artificial nipples. If they sucked at a certain rate a recording of their mother's voice reading a story would be activated. Sucking at a different rate switched on a different female voice reading the same story. Out of ten babies tested, eight quickly got the hang of calling up their own mother; indeed, in some cases it took only ten minutes for the baby to learn the sucking pattern. When the pattern was switched, the babies relearned the new pattern, changed their sucking rate and called up their mother. As the babies had had only a few hours contact with their mothers, being less than three days old, it was speculated that the mothers' voices had been heard while the babies were in the womb.

Pigs and Sheep

Elizabeth Walser, of the Institute of Animal Physiology at Cambridge, has been studying the vocal repertoires of domesticated animals and the role that sound plays in their lives.

Pigs have well-marked vocal communication. Piglets, for example, have a contact call that they make when a litter becomes separated. There are warning barks and greeting honks from adults, which are given to strangers or familiar farm workers approaching the pen or field. The boar has a definite sequence of grunts during courtship of the sow, a feature used in artificial insemination for testing the condition of the sow. A sow in heat will stand still when she hears the rhythmic grunting of the boar.

Much of the work at Cambridge is concerned with animal welfare and animal husbandry. Pigs, it seems, make happy and unhappy sounds, which can be monitored by placing a microphone in the pen. A pig will protest if another lies on top of it. And there is an asking call for food.

Lactation grunting has received particular attention. When a sow is about to feed her litter of piglets, she will lie on her side and call the youngsters to her. She will grunt, slowly at first. Each piglet finds its teat and rubs its nose around it. The grunting gradually gets faster. Just before it reaches its fastest rate, the piglets stop rubbing their noses and home onto the teat. Their ears stick up and they become tense. The sow lets the milk down and the piglets start to suckle.

Together with Lyn Jefferson, Elizabeth Walser decided to investigate whether the maternal grunting was synchronising the behaviour of the piglets. An artificial sow was made out of a large spongy udder with a hairy top on it. Tape recordings of the grunting were synchronised to the artificial feeding mechanism in order to simulate a natural feeding sequence. The tape recorder was switched on, the grunts woke the piglets up, and they came to the artificial sow. The little ones behaved just as if they were feeding from a normal sow. The milk came down at the right part of the lactation sound cycle, and the piglets suckled.

In the next experiment the sound and lactation mechanism were put out of step. It was found that, if the milk was let down before the fast grunting, the piglets coughed and choked because they were not ready to receive it. If the milk was delayed, the piglets became agitated and made little encouraging grunts themselves. The researchers concluded that, although the piglets might be using features like tension in the udder and teats as cues, they were undoubtedly using the grunting as a signal for the feeding sequence.

Piglets recognise the voice of their own mother. Given a choice of two artificial udders, piglets would always go to the one replaying the maternal grunts. This caused problems at first in the lactation experiments, for when the mother's voice was played behind the artificial udder, instead of racing towards the artificial sow, the piglets would make straight for their sleeping mother, much to her surprise! What it is in the call which the piglets recognise is not yet known.

For mother-baby recognition the research team have been looking at ewes and lambs. Elizabeth Walser first noticed that Soay sheep, the wild, undomesticated sheep from Scotland, were silent except during lambing time. Bleating, she thought, must have some influence on maternal behaviour. Walser investigated whether lambs could recognise their mother's voice, and whether the ewes could identify the voices of their lambs.

In one experiment Clun Forest, Finnish, Jacob, Dalesbred and Soay ewes were hidden behind canvas sheets. Lambs were given the choice of three ewes, but they could only hear the voices. Over three-quarters of lambs tested were able to find their mother by voice alone. The ability improved with age. Three weeks old lambs were far more proficient than the new-born lambs. Jacob lambs were not as good as the others.

In a second experiment ewes were invited to find their lambs. Surprisingly, fewer than half the ewes could find their lambs by voice alone. This, however, is consistent with the behaviour of sheep in the field. Ewes tend to stand still and bleat. The lambs run to them. Only under circumstances of extreme pressure will a ewe seek her lambs.

In a third experiment ewes were played tape recordings of lambs bleating. The ewes replied to recordings of their own lambs' voices more than to the bleats of alien lambs. The mothers tended to answer recordings of their own offspring more quickly. The experiment was taken further by using the lambs themselves instead of recordings, and Walser found that ewe and lamb seemed to be having a conversation. The lamb bleats, the ewe answers, and the lamb replies again, sometimes culminating in a very fast exchange of ewe and lamb vocalisations. The research team are attempting to analyse the sequence to see if the conversation will give some idea of how the ewe has recognised the lamb and the lamb has recognised the ewe.

Another surprise is that different breeds of sheep show quite different abilities in mother-offspring recognition. Dalesbred, hill-sheep lambs could find their mothers much more efficiently than lambs of other breeds. Black and white spotted Jacob lambs turned out to be the dunces at recognising mother by voice. Walser and her colleagues decided to investigate why there should be this discrepancy. Was it something in the ewe's voice that made it difficult for Jacob lambs to identify their mothers or was it something in the hearing ability of the lambs? To find the answer, embryo transfer techniques, developed at the Institute, were employed. Some Dalesbred embryos were transplanted to Jacob ewes and Jacob embryos transplanted to Dalesbred ewes, such that each ewe gave birth to a mixed set of twins. Eight Dalesbred ewes and eight Jacob ewes gave birth to twins, one a Dalesbred and one a Jacob lamb. The lambs were reared quite normally. The ewes treated the lambs as if they were a normal pair of twins. The lambs were tested for maternal voice recognition and it was found that all the lambs born to Dalesbred ewes could distinguish their mother's voice much better than all the lambs born to Jacob ewes. This suggested the difference was in the voice of the ewes. Dalesbred ewes are more understandable.

Walser analysed the ewe vocalisations on the sonograph, looking at the fundamental frequencies and the structure of the ewes' voices. She found that Dalesbred ewes have considerable individual variation in their voices. Jacob ewes have homogeneous voices. Walser also noticed that Jacob ewes are far more vocal. They bleat more readily if a youngster is taken away. Dalesbred ewes only answer the call of their own offspring.

Looking at the social grouping in flocks, the Cambridge researchers have found that new lambs stay close to the mother for some time. There are family groupings. Male juvenile lambs form male groups, and ewes and female juvenile lambs stay together in a separate group. Sibling pairs of lambs recognise and respond to each other's voices. About a week after birth a ewe will only feed a pair when they are together. The lambs learn very quickly that there must be two of them at the dinner table before they will get their milk. In the field, one lamb can often be seen looking for its twin exchanging bleats, and the two of them running to the mother. If one lamb turns up alone it is butted away until the other arrives. Twins tend to stay together in the field. In tests a lamb will prefer its sibling to an alien lamb.

The first sound a lamb hears when it is born is a low-frequency rumble, a purring sound coming from the ewe. The lamb and the ewe stand head-on to each other, while the ewe starts to lick the lamb's head. As she moves further down the body the rumbling sound appears to comfort the lamb. The sound is produced much more if the lamb is vocal and apparently agitated. In the dark the rumble seems to help the lamb to stay near the ewe. The sound does not travel far. Elizabeth Walser feels that it is a sound which can pass between mother and offspring without attracting the attention of predators.

8
DANGER

In *Behavioural Ecology: an evolutionary approach*, Paul Harvey and Paul Greenwood of Sussex University wrote: 'No single problem in anti-predator behaviour has attracted more theories and produced fewer facts than that of the evolution and function of alarm calls in birds and mammals'.

Bird Alarm Calls

The redshank *Tringa totanus* is often known as the 'sentinel of the salt-marsh'. It is first to spot danger and first to raise the alarm. The redshank's alarm call is shrill. Unlike other calls, the alarm call warns not only other redshanks but also birds of other species; curlew, plovers, gulls soon follow with a chorus of concern. In the mass confusion a predator is often distracted and maybe deprived of its meal. In the wood, the jay *Garrulus glandarius* is usually first to warn of an approaching threat. A day-flying owl will evoke a hard rattling call from the jay which serves to encourage other birds in the trees to join in and, together, noisily to chase away the predator. In the garden, the robin *Erithacus rubecula* gives a whispy call as a crow flies overhead. The sound may be untraceable but all the other birds nearby become agitated.

Alarm calls seem to fall into three types: those like that of the redshank are clear shrieks of despair which travel a fair distance across the marsh or estuary; those of the jay are hard-edged and raucous and draw the attention of neighbours to the area of excitement; and those of the robin travel short distances and are hard to pinpoint.

Alarm calls which are difficult to locate have been of particular interest to researchers. When at Cambridge with Thorpe, Peter Marler analysed the way chaffinches respond to danger. They have several alarm calls, two of which are the 'chink-chink' and the 'pink-pink' calls. For most other calls, the transmission range is maximised, but for these signals the opposite is required, or at least a compromise between communicating with individuals that need the message and those that would become a threat if they picked up the signal. The least locatable of chaffinch alarms is a call given by the male chaffinch to the mate and young at the nest when danger threatens. It is a very high, thin whistle which has the curious property of being extremely difficult for a human observer to locate. Peter Marler was curious about the quality of this call and consulted a colleague in the psychology department, Donald Broadbent. They carried out some simple experiments which showed that this alarm call is designed in such a way as to reduce the cues that might enable a potential predator to pin-point the calling bird. It is not completely non-locatable for it is not possible to produce such a sound, but it does minimise the cues available.

There are basically three kinds of cues that humans use in locating a sound, all involving some comparison between sounds reaching the two ears. The time of arrival of sounds is important; so, too, the intensity of the sounds. The head casts a sound shadow so if sound is coming from the left, the left ear will hear a louder sound than the right. The third feature is a difference of phase at each ear.

As sound is in waves, the timing of the peaks and troughs can be identified. To do this accurately the sounds must contain easily detectable abrupt discontinuities, the moment of arrival of which can be compared at a bird's two ears. Using that comparison the bird can locate a sound source. Unfortunately for the chaffinch, predators can do the same. The chaffinch gets around this by producing a call which gradually fades in, reaches its maximum and then slowly fades out. The high-pitched nature of the whistle also makes it difficult for comparisons of phase. Wavelengths shorter than the distance between the ears would mean that a receiver could not be sure if it was hearing the dip of the trough directly after the first cycle peak that it heard, or whether another cycle or two had intervened. Peter Marler suggested this explanation for the nature of the chaffinch 'hawk alarm' in 1955. Since then the story has become more complicated. Bird predators, for example, may not be carrying out these kinds of comparisons. The way they localise sounds may rely less on a pressure-receptor system than on a pressure-gradient or particle displacement system. In this way a bird could locate a sound source simply by scanning with one ear. It has been shown that the diurnal goshawk *Accipiter gentilis* and the nocturnal barn owl *Tyto alba* can accurately locate birds emitting thin alarm calls. Michael Shalter and W. M. Schleidt, of Ruhr Universität, demonstrated that barn owls can locate clay-coloured thrushes *Turdus grayi* by their 'sect' alarm call.

However, some of these properties must be important, for thin, high-pitched, fade-in-fade-out alarm calls have been evolved independently by a host of other bird species. Not only that, but the call says meaningful things to birds of different species. Alarm calls are unusual in communicating across species boundaries. A chaffinch in the top of a tree might spot a sparrowhawk, give a hawk alarm call, which would transmit the presence of danger to a great tit, which would in turn signal to a robin and so on. Inexperienced bird-watchers often confuse the different calls. The 'wheet' call of the chaffinch,

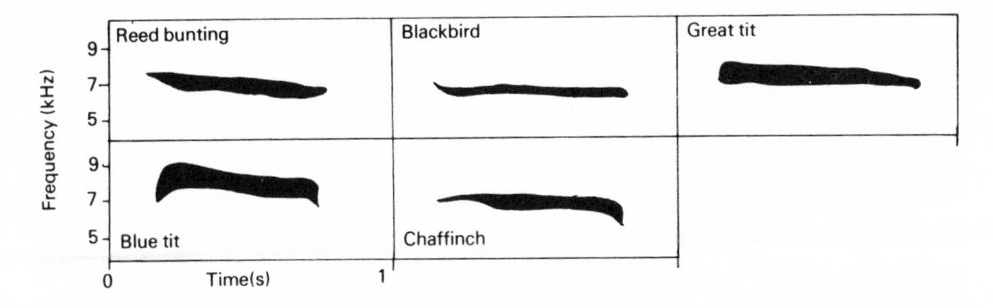

Fig. 26 Alarm calls to predator flying overhead.

for example, is almost identical to the anxiety call of the chiffchaff *Phylloscopus collybite*, the willow warbler *P. trochilus*, and the redstart *Phoenicurus phoenicurus*.

Peter Grieg-Smith, now at the Worplesdon Laboratory of the Ministry of Agriculture, Fisheries and Food, but previously at Sussex University, has been listening to the alarm calls of stonechats and has discovered a two-tiered defence system – one difficult to pin down like that of the chaffinch, but the other surprisingly easy to locate. Throughout the breeding season the stonechat has a characteristic call with two different sets of notes. One note is a thin, short whistled sound which has been dubbed the 'whit' note. The second consists of harsher 'chuck' notes which sound a little like two stones being struck together and which may have given rise to the name, stonechat.

By watching stonechats in the field, Peter Grieg-Smith showed that the two types of calls are related to the type of predator. Stonechats face two different threats. From the air birds of prey may attack the adults themselves and their offspring. Terrestrial predators are a threat only to young in the nest. A sparrowhawk approaching provokes the 'whit' call from a parent, whereas a human or a dog are greeted with the 'whit' and 'chat' calls mixed together and delivered in a seemingly frantic string of loud notes. Peter Grieg-Smith felt that the 'whit' call is a warning call. It fits Peter Marler's hypothesis that such a call should be thin, covering a narrow band of high frequencies and fade in and fade out, thus making it difficult to locate.

To test for this Peter Grieg-Smith took nestlings into the laboratory and measured the way in which they begged for food while he played tape recordings of different sounds in the background. When adult 'whit' calls were played the nestlings virtually stopped begging. 'Chat' calls were ignored. The 'whit' call must give the nestlings some kind of warning of danger which causes them to stay still and quiet. Peter Grieg-Smith used himself as the 'predator' in trying to evaluate the 'chat' call. As he neared a stonechat nest the adults called loudly with 'whit' and 'chat' calls and moved around the territory trying to draw him away from the nest. He noticed that if he moved closer to the nest the rate of 'chat' calls went up and if he moved away the rate was reduced. The calls were accompanied, inevitably, by wing flicking which exposed flashes of white from the conspicuous wing coverts. The 'chat' call, he concluded, is a distraction display. It is easy to locate, being hard-edged and raucous.

Grieg-Smith also looked at variations in nest defence behaviour throughout the breeding season and noted that peaks of activity occurred just after chicks had hatched. Stonechats often have three breeding periods each summer and the rates of calling increase each time after hatching. Grieg-Smith suggested that increased begging activity with louder, noisier calling by the chicks would require more 'whit' calls from the parents to keep them quiet and not attract predators to the nest.

The 'whit' and 'chat' calls, together with visual distraction displays, have evolved as a means of protecting the stonechat's offspring. But the young stonechats were not alone; other birds in the heathland ecosystem have realised that listening to the stonechat's calling might have benefits for them too. Stonechats spend a great deal of time on exposed perches from which they can spot danger from afar. Linnets, redpolls, yellowhammers, and reed buntings spend much of their time in the bushes, hidden from predators but also unable to see them coming. By associating with stonechats these birds have discovered that the stonechat alarm system is an effective early warning system. It is a kind of insurance policy which, in addition, allows them more time for foraging with less time spent on scanning the skies for danger. Grieg-Smith compared meadow pipits *Anthus pratensis* foraging alone with those foraging with stonechats. Pipits in the company of stonechats spent half as much time in predator surveillance as unaccompanied ones. Stonechat alarm calls, like those of the chaffinch and many other birds, serve to warn, not only individuals of their own species but also those of other species.

Steve Wilkin and Millicent Ficken, of the University of Wisconsin at Milwaukee, studied the alarm vocalisations of the black-capped chickadee *Parus atricapillus*. When a predator approaches a flock of these birds, an alarm call, known as 'High Zees', is emitted. The call fits with Peter Marler's hypothesis that it should minimise directional localisation by gradually fading in and out, and it also fits Eugene Morton's hypothesis that it should be a signal strongly attenuated by the environment by having a high pure-tone frequency of about 7,940 Hz. It is also given at an intensity which would travel to other members of the flock but no further, so that it would not reach a predator. It is also possible, because of the high frequency of the signal, to beam the sound to a receiver, and away from the predator.

Wilkin and Ficken suggested that the evolution of the chickadee's alarm call was *not* related to ensuring the safety of relatives and offspring as youngsters often disperse many miles away to other flocks. They also concluded that it was *not* to panic the flock and cause confusion for the predator as chickadees tend to freeze on hearing the alarm call. Rather, the clue seemed to be in the 'long-term monogamous mating system'. Mated pairs of chickadees tend to stay close to each other in the same flock and alarm calls might be directed at a mate, to protect it from predation. In this way a pair of successful breeding birds try to ensure they survive together until the next breeding season.

The Owl

For one group of predators ventriloquial alarm calls do not help the prey. Owls can locate the whispiest of sounds. Owls mainly hunt at night, so coupled with their keen night vision capability is a remarkable sound location system. It has been said that of all animals tested for their ability to localise sound sources, the owl is the most accurate. In addition, it can localise sounds in both the horizontal and vertical dimensions. A sophisticated filtering and selection system to the brain, processing the sound signals heard by the bird, allows the barn owl *Tyto alba*, for instance, to locate its prey even in total darkness.

As an aerial predator the owl must be able to evaluate the position of its prey from above. In the case of the barn owl the target is often a small mammal. In about 95% of cases, where pellets have been examined, field mice have been identified as the main dietary constituent. Mice are quite noisy as they go about their nocturnal foraging. They rustle about as they scuttle through their runways in the undergrowth. They often squabble, squeaking away in both the audible and ultrasonic frequencies. It is these high frequency components of a rustle or a squeak that the owl uses to locate its prey with accuracy.

It might be expected that high frequencies would travel only for short distances, attenuating faster in air than lower frequencies. Also, high, hard-edged sounds bounce more readily off grass stems and twigs which should make them difficult for the owl to detect. The owl, however, has many tricks up its sleeve, or rather in its face.

The barn owl's face is basically a disc with two concave troughs which serve as a stereo parabolic reflector. It can focus high frequency sounds, particularly those between 3,000 Hz and 9,000 Hz. The facial ruff, as it is known, is made up of rows of tightly packed feathers dividing the face into two ear-like depressions. The depressions function much like the external pinnae of human ears in channelling sound to the ear proper. The ears of the owl are behind the eyes, at the focus of the two facial depressions.

Furthermore, the ears are not placed symmetrically on the head. The right ear is directed slightly upward, and is more sensitive to sounds coming from above, whereas the left ear is focused downwards and responds better to sounds from below. With this sophisticated sound-receiving system the barn owl can compare timing and loudness differences at the two ears, not only from left and right, but also from above and below. So the owl is capable of detecting faint sounds, like mouse rustlings, at long distances, and of pin-pointing the sound source to within a degree (the equivalent of a little finger width at arm's length).

The way an owl uses its facial ruff to home-in on a target has been observed in the long-eared owl *Asio otus*. At rest a long-eared owl collapses the ruff. If it

Fig. 27 Facial structure of the barn owl. (After Konishi)

is startled the ruff is expanded and the head turned to face the sound source. The 'ears' of the long-eared owl, incidentally, are not sound receivers but are probably structures for simulating the interspecies threat displays of other predators. 'Horned' or 'eared' owls and mammalian predators often meet. Ivor Mysterund and Henning Dunker, from the University of Oslo, suggest that eagle owls and lynx, for example, both predators in northern Scandinavia, are likely to have 'face-to-face' encounters. Eagle owls' nests are accessible to lynx, foxes and martens. By adopting a 'lynx-like' face complete with ears, the owl is thought to deter these mammals.

The owl's auditory sophistication was first recognised when Roger Payne, then at Harvard, noticed that, as an owl swoops down, it orientates itself with respect to the target, whichever way the prey runs. The owl aligns its talons with the body axis of the prey; a remarkable feat if you consider the unpredictable zig-zag running of, say, a fleeing field mouse. The owl must change the alignment of its talons rapidly with every manoeuvre. Experiments in the laboratory showed that the owl could do this in complete darkness; it must therefore rely on sound cues.

To test for this, Masakazu Konishi and Eric Knudsen carried out a series of head-orientation experiments. Observers had noted that an owl, on locating a sound source, quickly turns its head towards the sound so bringing it directly in front of the face. This directs both eyes and ears on the target. Konishi and Knudsen set out to measure how accurately the owl could turn its head towards a sudden sound. To do this they placed an owl inside an electric field produced by induction coils, placed a small search coil on the owl's head, and measured the electrical changes that took place between the search coil and the large induction coils as the owl moved its head. They could then accurately measure any movement the owl made. Two sound sources were used. The first, the 'zeroing' loudspeaker, was placed directly in front of the owl. The second, a mobile 'target' loudspeaker on a track, could be moved around the owl. It could also be moved up and down. Tests were carried out in the dark. A sound would be put through the static 'zeroing' loudspeaker and the owl would turn its head to face it. Then, a sound was put through the 'target' loudspeaker, the owl would rapidly turn its head to face the speaker, and the time taken and accuracy of the movement noted. The whole system was motorised and computer controlled.

Using this ingenious apparatus, Konishi and Knudsen were able to show that an owl's head would orientate towards a sound source even when the signal was terminated before the head-turn began. This showed that the owl had registered vertical and horizontal information and determined the location of the sound source before even turning its head. To accomplish this the

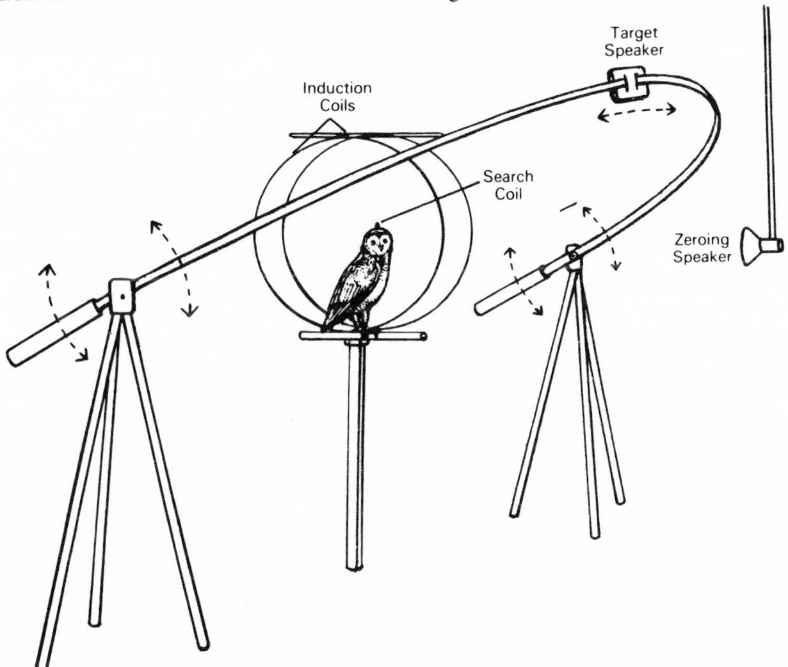

Fig. 28 Head orientation and sound location tests with the barn owl. (After Konishi)

owl makes a comparison of the timing, intensity and quality of the sounds reaching its two ears, and tests have shown that it is primarily interested in the intensity and directional components of the signal. Because of the size of the owl's head and the separation of the ears the phase delay of high frequencies only is registered. Konishi has shown that owls are able to locate sound sources on the floor of a large anechoic chamber by hearing loudspeakers under small trap-doors in the floor. Owls found high pitched calls easier to locate than low pitched sounds.

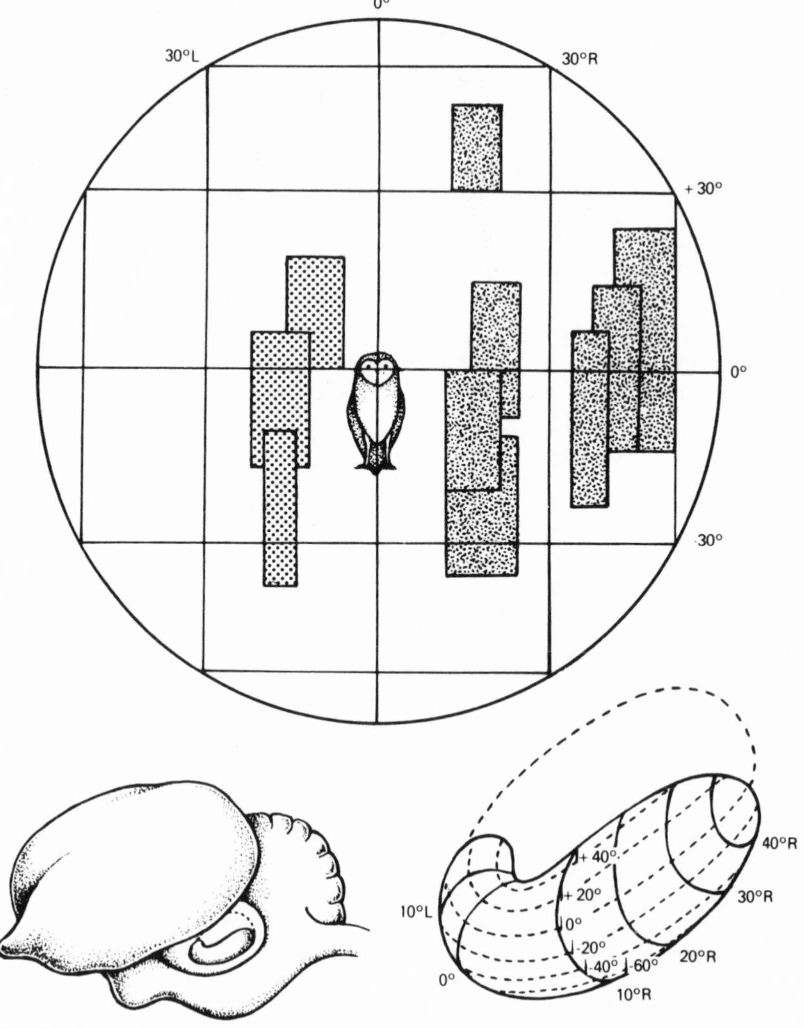

Fig. 29 Two-dimensional map of space found on the left side in the mid-brain (B) of the barn owl. Each neuron responds to a specific region of space (A) mainly on the right side and is arranged in receptive areas that follow spatial contours (C). (After Konishi)

Sound signals arriving at the ears are converted into nerve impulses which are relayed to the brain. Here the owl has an amazing way of processing the information. It creates a sound map of the world about it. Each neuron is only excited by sound signals coming from a particular point in space. The sound map is constructed in the midbrain – the mesencephalicus lateralis pars dorsalis – and the information sent to an area of the cerebellum called Field L (equivalent to our own auditory cortex). The experimental work continues at the Los Angeles laboratory to determine how neighbouring cells in the owl's midbrain are capable of responding to very subtle spatial differences in the sound source.

Mobbing

If an owl corners an unfortunate victim, all is not lost for many animals have another noisy defence system to bring into action – mobbing. Owls approaching a nest or feeding area in the daytime will be mobbed by many of the smaller birds in the area. Thrushes, jays, nuthatches and chaffinches, for instance, make mock attacks, swooping in on the intruder, retreating, circling and attacking once more. Characteristic harsh calls of alarm are emitted during the harassment. Like the 'chat' call of the stonechat these mobbing calls are easy to locate and serve to gather potential defenders from the surrounding countryside. Mobbing may work in a variety of ways. The noisy attacks may inform others that danger is threatening. They might tell the predator itself that it has been spotted and that there is no chance of a surprise attack. Mobbing might also drive the predator away so that it hunts elsewhere. A more cynical interpreter might consider that a bird producing raucous mobbing and alarm calls is attracting as many other birds to itself as possible, so creating an artificial 'selfish-herd' situation. By manipulating other birds to congregate around a predator, an individual will reduce its chances of being caught.

Mobbing calls are short, 'square-edged', and have a wide range of frequencies present. This helps orientation for others as the broad frequency spec-

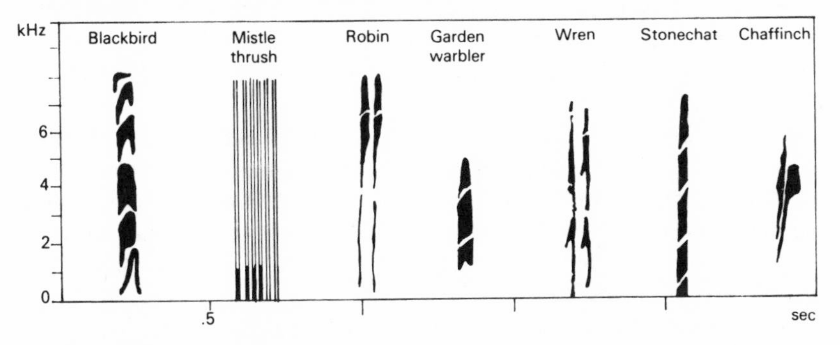

Fig. 30 Mobbing calls given to an owl.

trum gives phase and intensity difference information to a receiver and the 'square-edged' starts and stops provide time difference cues.

One bird which has been well studied for its anti-predator behaviour is the pied flycatcher *Ficedula hypoleuca.* At Ruhr Universität, Bochum, Eberhard Curio noted two ways in which flycatchers deter predators. In response to birds, such as great spotted woodpeckers, likely to threaten eggs and nestlings, the flycatchers carry out close attacks, known as 'snarling'. Mobbing from a distance is reserved for sparrowhawks and other predators dangerous to adult birds. Before the breeding season begins, mobbing responses are at a low level. They increase when the eggs are laid and when the young hatch, reaching a peak just prior to fledging. If a bird loses its brood, mobbing is reduced. Unmated birds expose themselves much less to threatening predators than mated pairs, suggesting that mobbing behaviour, in this species, has developed to protect offspring although there is some risk to the parents. Curiously, Michael Shalter showed that this mobbing response by pied flycatchers can be reduced if the birds become habituated to a predator (in this case a stuffed owl) always attacking from the same direction; the mobbing response is only evoked when the predator is presented from another direction. Shalter also demonstrated that birds of different species will orientate towards the source of a mobbing call, including the predator itself.

One other function of mobbing has been revealed by Willy Vieth and Ulrich Ernst, working with Curio at Bochum. In a series of experiments with European blackbirds *Turdus merula* the researchers found that mobbing behaviour and its associated production of mobbing calls may facilitate 'cultural transmission of enemy recognition'. A blackbird exposed to a danger stimulus without having experienced mobbing will show a weak response to the danger. If it is able to watch and listen to mobbing behaviour by birds of its own species, then the next time it is exposed to the stimulus it mobs more strongly than it did initially. Curio and his colleagues were able to show that mobbing behaviour in an individual blackbird could be enhanced simply by tutoring it with mobbing calls alone, played through a loudspeaker each time a danger stimulus was presented. In addition, the mobbing calls of species other than blackbirds were almost as effective in tutoring the test bird, although the strongest mobbing response was seen when an individual was presented with a stimulus and provided with both the visual and acoustic mobbing behaviour of other blackbirds.

Mobbing behaviour, it seems, is related to the protection of the offspring and may function by warning and teaching about danger. It is not confined to birds. Chimpanzees and baboons will mob leopards. Chimps will reinforce their cries and screams by standing upright, brandishing sticks and throwing rocks. Lemurs and ground squirrels mob snakes and axis deer have been known to harass tigers.

Owls, though, have taken advantage of mobbing behaviour. There have been several reports of a male owl flying noisily down a woodland clearing, attracting the attention of all the small birds in the area. They respond by mobbing him. The female meanwhile has been waiting and watching in a nearby tree. As the small birds congregate she swoops down and grabs one.

Escape

Often auditory signals help make escape, another form of defensive behaviour, more effective. Crickets, for instance, stop singing when danger threatens. Screech beetles screech. By producing a loud noise the insect might startle a predator momentarily and maybe escape. Frogs do the same. The marsh frog has a clearly audible squeak as it leaps back into the safety of the pond. If a frog is picked up it will often surprise the handler by emitting a sudden high-pitched scream. Göran Högstedt, of the University of Lund has suggested that screams attract other *predators* looking for an easy kill and that the unfortunate victim is able to escape in the dispute between its original captor and the new arrivals.

Many small mammals, like ground squirrels, rabbits and prairie marmots, escape to burrows. Usually an alarm call warns a colony of an invader. Paul Sherman, now at Cornell University, studied Belding's ground squirrels *Spermophilus beldingi*, diurnal, group-living rodents that live in the Sierra Nevada mountains of California. At 3,040 metres on the summit of Tioga Pass, Mono County, California, Sherman watched groups of squirrels in a meadow. Ground squirrels hibernate for most of the year, becoming active between May and October. A female will rear one litter a year and, if she loses that litter through predation, indulges in a curious bout of infanticide. She will move to a new area, carefully chosen to be less susceptible to predators, and then go about the wholesale slaughter of the infants of any females in the area. In their family groups, however, they are less violent, each group looking after its relatives. This is shown particularly with alarm calls. A mother will put herself at risk by calling to relatives when danger threatens. An individual without a family will not call. Sherman proposes that these squirrels are showing altruistic behaviour.

To aerial predators ground squirrels give single note, high pitched whistles. To mammalian predators the call is segmented and falls between 4,000 and 6,000 Hz. The caller is easily spotted because of the vigorous vibrations of the chest cavity and the open mouth.

At Burns, Oregon, Scott Robinson, of the University of Wisconsin, Madison, watched another population of Belding's ground squirrels and observed their responses to different predators. Clearly it is to an animal's advantage if it distinguishes between dangerous and harmless situations, saving on valuable time and energy. Robinson found that ground squirrels are able to 'discriminate between predators of squirrels and harmless animals in both aerial and terrestrial encounters'. Swainson's hawks, prairie falcons, northern harriers, black-billed magpies and ospreys elicited alarm calls and escape manoeuvres.

White pelicans and turkey vultures, on the other hand, may be greeted with chirps but the animals would not be put to flight. Dangers on the ground come from large mammals, like dogs, coyote, mink and badger, small mammals, like long-tailed weasels, and snakes, like the gopher and rattlesnake. Squirrels responded most to the large mammalian predators with calls, posting behaviour (getting onto the hind legs with the body upright), and escape.

Robinson found also that ground squirrels could 'discriminate many different groups of predators'. Harriers were separated from other raptors and omnivorous birds. Large mammalian predators were received differently from weasels, snakes and other ground attackers. They could, in addition, 'discriminate among contextually different situations with the same kind of predator'. Harriers flying overhead were considered more of a danger than those moving south on the other side of a canal. Egrets in the air were considerably less dangerous than those approaching on the ground.

Depending on the predator, the squirrels responded in different ways. Chirps and trills and posting behaviour usually meant very dangerous animals were about. This might be followed by the animals fleeing, sometimes on the surface but often by taking evasive action by running through their tunnels. On some occasions the squirrels might turn about and chase the danger away.

The fleeing response to different terrestrial predators is interesting. Alarm calls given to a coyote or domestic dog seem to say 'it's OK to dive down any burrow to escape', whereas when a badger is about the cry is 'find a burrow with a back-door'.

Wilma George, at Oxford University, has studied another charming little creature which has a small vocabulary of alarm calls – the gundi *Ctenodactylus gundi*. Gundis are small rodents that look a little like guinea pigs, and they live in the rocky parts of African deserts north of the equator. They are noisy

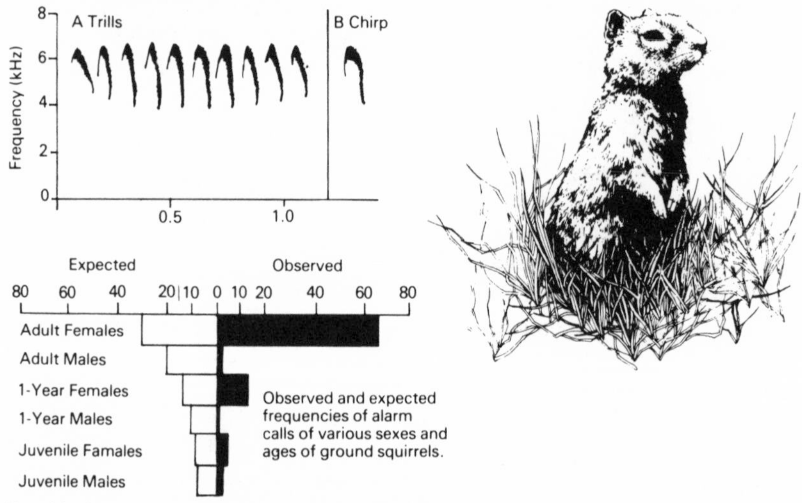

Fig. 31 Alarm calls of Belding's ground squirrels.

Fig. 32 Gundis.

animals, with a basic alert call which is used in greeting, in squabbles over food, and when something interesting has been spotted. The alert call is relatively long and frequency modulated, almost a bird-like chirp. There is a low frequency component (1,200 to 2,200 Hz) followed by a high frequency component (4,600 to 6,000 Hz). It is often used when a predator is a safe distance away, and the gundi is edging its way back to its tunnel entrance ready to escape if necessary. When the predator comes nearer the call is shortened. A panic call is given if a shadow or aerial predator passes over.

Another species of gundi, the desert gundi *C. vali* has an alert call which is even more bird-like. Clear, fluted rising and falling syllables are repeated typically four times per phrase in the frequency range 2,000 to 4,300 Hz.

It is thought that a gundi's vocal repertoire, unlike the ultrasonic calls of many other rodents, is pitched relatively low as only low frequency sounds are not absorbed by the hot dry air of the desert.

These little creatures also have large ear bullae, and thus enlarged middle ear apparatus, an adaptation for increasing hearing ability when desert conditions would normally favour the development of small ear pinnae on the outer ear. This was noticed in many desert rodents by H. Heim de Balsac in 1936. Later it was suggested that this adaptation is for picking up low frequency sounds and might be related to the detection of approaching predators.

9

THE FIRST VOICE

About 350 million years ago, in the steaming primaeval swamps, a new group of vertebrate animals was beginning to evolve the abilities which would enable them to live on land – they were to become the amphibians. At that time, fish, crabs, lobsters and shrimps living below the water, and insects and spiders already on land, were making their instrumental scrapings and raspings. But, some 200 million years ago, the first *voice* in the history of the earth was probably that of an amphibian – maybe something like that of a frog.

Frogs, it has been said, divide their world into three classes of objects – if it's small enough you eat it; if it's too big you run away from it; and if it's intermediate you mate with it. In order that frogs mate with the right object, namely a female of the same species, they rely on sound. There are frogs and toads that chirp, others croak; some whistle; and there are those that click. Some species produce constant pure tones; others very noisy broad-band calls. Two species of *Kassina* frogs from South Africa, the predominantly terrestrial *K. senegalensis* and the aquatic *K. maculata* have frequency modulated calls. *K. maculata* emits a mating call that sweeps upward in frequency from about 800 Hz to over 3,000 Hz in approximately 15 milliseconds, a feat to rival sound production in birds and bats. Indeed, the sweep is very much like the modulated sweep of a bat except that it lies in the range of audible frequencies.

The Auditory System of Frogs and Toads

Frogs make their sounds by forcing air from the lungs across vocal chords, into the buccal cavity. By keeping the mouth closed and shutting the external nares the air can be shunted back and forth across the vocal chords to produce the characteristic chirping sounds. These can be further enhanced with the aid of a single vocal sac below the chin or a pair of balloon-like sacs on either side of the mouth, which serve to amplify the sound. A natterjack toad *Bufo calamita*, with the aid of its gular pouch, can be heard two kilometres away, whereas a common toad *Bufo bufo*, without a vocal sac, reaches to only 150 metres.

Frogs croak to attract females. To avoid being spotted by predators, most frogs and toads mate at night when it would be ineffective to leap around or perform some visual display. So the best way to attract attention is by a sound signal. One of the pioneers of research into the calls of frogs and toads is Dr Bob Capranica from Cornell University, New York. His aim has been to learn something about the role that sounds play in the lives of frogs and toads and then to discover how they locate and recognise those sounds.

Capranica started at the University of California, Berkeley, graduating in electrical engineering. He became interested in communication theory and went to Bell Telephone Laboratories, designing electronic communication systems. With a Bell Labs Fellowship he was able to go to the Massachusetts Institute of Technology where he met biologists and medical researchers interested in speech. He decided to pursue studies of communication in animals. MIT's novel analytical tools were to change the nature of this field of research. No-one at that time was approaching studies of animal communication at the levels of sophistication at which scientists were pursuing, for example, studies of speech. Ken Stevens had built the first computer to synthesise speech sounds, and Capranica thought that these techniques might be useful in studying animal sounds. His first study animal was the bullfrog *Rana catesbeiana*, the largest of the North American frogs. He built indoor terraria, and brought frogs into the laboratory. At first, Capranica noticed that occasionally a male bullfrog would call with a deep, resonant croak and another male in the laboratory would be triggered to call. Capranica called this the 'evoked calling response'. He carried out a series of simple playback experiments, exposing male frogs to the recorded call of others. The males always called back. They wouldn't respond, though, to the recorded calls of other species of frogs or toads. The vocalisations of bullfrogs, at least, Capranica concluded, are species specific.

There is also the suggestion that male bullfrogs recognise individuals by differences in their calls. Males are territorial and fight, initially with neighbours, to establish their boundaries around the choice territories. Bullfrogs prefer to spawn in areas of the pond that are least subject to extreme variations in temperature, which might cause abnormalities to appear in the developing embryos, and that are least likely to harbour embryo-eating predatory leeches. Females appear to choose males on the basis of the quality of the territory they have acquired. Older and larger frogs muscle-in on the best territories. After a period of vocal contest with neighbours, when demarcation lines are established, neighbouring male bullfrogs tend to get used to each other's calls and only act aggressively towards intruding strangers. If a neighbour's call is played to a resident male, generally it will be ignored, the resident continuing to call to attract a mate. If a stranger's call is played the resident will stop singing and rush towards and threaten the loudspeaker.

These calls, however, are not the only sounds emitted by bullfrogs. Capranica identified other different vocal signals such as an alarm call, a distress call, calls associated with feeding behaviour, and the release call. The distress call is made by a bullfrog when it is being attacked by a predator. It is a very high-pitched scream, lasting several seconds, a little like a baby's cry. The scream is a startling sound, the only call made with the mouth wide open. If a bullfrog is grabbed by a racoon, or some other predator, the scream can be heard across the pond. Whether this is to warn others of danger or an attempt to startle the predator is not clear. The other interesting sound made by bullfrogs is the release call. Male bullfrogs are not very good at establishing the sex of a silent frog they may encounter on a dark night. If another frog should stray into the male's immediate vicinity and touch him, the calling male will stop calling and clasp the intruder in the hope that it is a female. If it *is* the right sex the female remains silent and the male retains his hold. If, however, it is an unreceptive female, maybe one already having spawned, or a male, then the animal will produce a distinct release call. This lasts for several seconds. The male recognises an inappropriate partnership and the animal is released. At the American Museum of Natural History, New York, researchers carried out a rather entertaining experiment which verified the significance of the release call. They took several balloons and partly filled them with air and water so they would float just below the water surface. A string was tied at one end and the balloon dragged along near to a calling male frog. The male stopped his calling and clasped the balloon. Since it couldn't produce a release call, the male sat there holding the balloon for the entire evening! The experiment clearly showed that the release call is necessary for discrimination of an appropriate receptive partner. Similar observations have been made in Britain with common toads. They will hold on tightly to any small moving object that does not emit a release call. They have been seen to attempt mating with goldfish and even Wellington boots!

Capranica recorded and analysed all the bullfrog calls and then, using the same techniques as the speech researchers, proceeded to synthesise them. Thus he was able to vary different parameters in order to find out which components of the sounds the bullfrog recognised as characteristic of its own species.

On returning to Bell Telephone Laboratories, after completing his doctorate at MIT, Capranica met a researcher from the University of Haifa, Israel, Eviatar Nevo. Nevo claimed that he had detected, simply by ear, geographic variations in the mating calls of cricket frogs *Acris spp.* Nevo and Capranica spent three years establishing that such variations existed. They made over 5,000 recordings and took them to Bell Labs for analysis. Then they started a series of playback experiments.

Two species of cricket frog occur in the USA, *A. crepitans* is found throughout most of the USA east of the Rockies, and *A. gryllus* is located mainly in the south-east.

During the spring, cricket frogs emerge from hibernation and congregate at their breeding ponds. The males begin to call. Populations of frogs, with many species present, can sometimes be quite dense so each male must compete with many others for his signal to be audible. The female enters the pond and selects a calling male of her own species. She may sit just six to eight inches away from him, listening to his repetitive calls, sometimes staying for an hour or two. Finally, when she has made up her mind that this is the frog for her, she swims the last few inches making physical contact with the male. He immediately stops calling and clasps the female in amplexus for up to three hours. As is the usual pattern in frogs, the female spawns and the male fertilises her eggs externally. All this happens on one night of the year only. The female is only responsive to a male's calls on that one night. Therefore, in order to test the female's discriminating abilities, Nevo and Capranica searched ponds and lakes for pairs of cricket frogs in amplexus. Clearly, the female is likely to respond more readily to playbacks at this time as her ovulatory cycle is in full swing, and she will spawn in two or three hours. If separated from the male at this point there is tremendous pressure for her to find a mate immediately.

Separated females were brought to an arena away from the calling males in the pond and placed between two loudspeakers on the ground. Through one loudspeaker was played the mating call of a male from the female's own area, and through the other the call of a male from a different area. The female was released and hopped immediately towards one of the speakers – sometimes scratching at the loudspeaker in order to get at the male presumably calling from inside. Using both natural and synthesised sounds, Nevo and Capranica showed that female cricket frogs would respond to the calls of males of their own local population rather than those of males of different geographical populations. Females in central Texas, for example, would react to the calls of males from central Texas, but not to the males of east or west Texas, or Alabama, or Louisiana. Cricket frogs, it seems, have geographical dialects.

By analysing the calls, the researchers identified a stereotyped pattern of clicks which is distinct, and recognised by the female, from each dialect area. Nevo and Capranica used sound to compare the evolution of both species and looked at how the characteristics of the sound vary geographically in different directions. In this way they could attempt to pinpoint the original location of each species. After analysis of 5,000 calls, one of the most complicated analyses of animal sounds so far undertaken, it turned out that cricket frogs started out in a region around Georgia, in the south-eastern part of the USA and radiated from this local focus. *A. crepitans* spread throughout most of the USA, whereas *A. gryllus* moved down through Florida.

Furthermore, the analysis included such environmental factors as climate, humidity, average rainfall, temperature etc. and it emerged that geographical variation in the calls can be linked with environmental factors. The sounds that each local population makes are adapted to the environment in which the animals live. Since certain frequencies or patterns of calls travel better under particular circumstances, the animals have evolved calls that maximise for long-distance communication.

Human ears are far apart, and a large solid object between our ears, the head, helps to provide a sound shadow. We assess the differences in intensity between the two ears, and identify minute time differences in the arrival of sound waves at each ear. But what about frogs? How does such a small creature localise sounds? This was a puzzle as the cricket frogs are about two centimetres in length from snout to vent, weigh just one gram, and the distance between their ears is less than a centimetre. In an attempt to find an answer, Capranica turned his attention to the mating calls of another North American frog. Living alongside the cricket frogs in Georgia, Alabama, and Florida is the green tree frog *Hyla cinarea.* Together with Carl Gerhardt, of the University of Missouri, and Jürgen Rheinlaender, of Ruhr Universität, West Germany, Capranica attempted to identify the features of the call that a female recognised, how well she localised the call, and what it is in the calls that permits localisation. Again, pairs of green tree frogs in amplexus were found. The female was separated, and mating calls played back to her through a movable loudspeaker. The researchers observed her approach to the loudspeaker with a video camera, and under very dim light. The arena was criss-crossed with grid-marks so that her trajectory could be traced very accurately. To their amazement, the research team found that a female green tree frog would orientate and hop towards the loudspeaker with an accuracy of about 10°, that is, she could pinpoint the source of a male's call to within an arc of less than 10°, and from four metres away. For such a small creature to locate with this accuracy is quite remarkable. Humans can localise sounds directly in front with an accuracy of about 3°, and sounds to the side to about 7°. The green tree frog is a tiny little animal with ear drums exposed directly to the air, flush with the head (an adaptation for streamlined swimming), and very close together. In addition, the sounds being used are relatively low frequency sounds with long wavelengths.

The researchers continued their experiments using a female which had already proved her skills by locating the loudspeaker to within 10°. She was re-released into the arena, only this time with grease placed on one eardrum. This, in effect, temporarily deafened that ear. The female became disorientated and hopped around in circles. If the left ear was covered by grease she would circle to the right; if the grease was on the right ear she would circle to the left. As she circled, she would continually wipe the side of her head with each leg in an effort to dislodge whatever was causing her hearing impairment. When the grease was removed and the female placed yet again in the arena, she would once more go directly towards the sound. Rheinlaender, Gerhardt and Capranica had verified, for the first time, that small vertebrate animals, like frogs and toads, require both ears and must carry out some binaural comparison somewhere in the nervous system; all the more remarkable, for the time of arrival differences at the two ears separated by less than a centimetre is only of the order of one or two microseconds, and the intensity difference a very small fraction of a decibel. It is hard to understand how the nervous system of such a simple creature could process such incredibly small cues, yet it appears that frogs can do it. How, in detail, they do it remains unsolved. It has been suggested that there is a pressure gradient system operating. Sound strikes one eardrum causing it to vibrate. Some sound goes through the eardrum, out the other side, along the Eustachian tube, and vibrates the other eardrum from the inside. Sound from both inside and outside reaches each eardrum. The sound on the outside compared with the sound on the inside has a different phase relationship because of the additional path length that the sound has to travel inside the tubes. As a sound source is moved around the frog in the external world, so too does the phase difference across each eardrum change. In addition, some sound is thought to enter through the nose, and travel via the mouth cavity, along the Eustachian tubes to strike the inside of the eardrum. So sound is coupled, not only from one eardrum to the other, but also through the nose. By using this clever technique small animals are able to achieve the same directional sensitivity as larger animals, without the external pinnae and a large head. More recent work by Gerhardt and Rheinlaender has demonstrated that female green tree frogs can also localise *elevated* sounds by scanning with the head, sometimes with the jaw parallel with the ground and sometimes with the head lifted.

The eardrum itself, though, is very large indeed. A typical male bullfrog, for example, has for its size one of the largest eardrums of all vertebrates with a diameter of over two centimetres. The female's is about half that size. This sexual dimorphism is only present in a few species; more normally the two sexes have the same size eardrums.

By bouncing a laser beam off the male bullfrog's eardrum, Capranica and his colleagues have shown that the amplitude of vibration is also very large. A sound of 90 dB causes the membrane to vibrate with an amplitude of 15,000 Angstroms. This is so large a movement that it can be observed through a light microscope. For comparison, a cat's eardrum, exposed to the same intensity of sound, vibrates with an amplitude of only 400 Angstroms. Why a bullfrog should have large eardrums that vibrate so violently is not clear.

The human ear has a threshold sensitivity which is more or less at the optimal theoretical limit. If it was more sensitive we would hear the molecules randomly striking the eardrum. Our auditory threshold sensitivity is, by definition, zero decibels. If our ears were ten decibels more sensitive, this background noise would interfere with communication and our ability to detect sounds at lower frequencies. Bullfrogs, despite their enormous eardrums vibrating at large amplitudes, have a low sensitivity of 30 decibels. Other frogs and toads are worse. The American toad *Bufo terrestris americanus*, which produces a trill 1,500 Hz call, has an auditory threshold of 50 decibels, and green tree frogs are even poorer. To make up for this, however, male frogs and toads make extremely loud sounds. A male cricket frog one metre away can be heard to be pumping out a painful 115 decibels. Some researchers have suggested that the rather dismal auditory threshold of many species may indicate that frog calls are mainly for short-range communication rather than signals designed to attract females to breeding ponds from a great distance. In fact, several studies suggest that females can locate suitable breeding ponds in the absence of calling males, perhaps by smell.

Another aspect of Bob Capranica's work has involved the way sounds are processed and recognised in the nervous system. What is it that allows a female to recognise a call from a male of her own species even when there is a cacophony of background noise masking the signal?

In order to attempt an understanding of the auditory processes, Capranica and his co-workers have developed apparatus which enables them to record from single nerve fibres in the auditory nervous systems of frogs and toads. An electrode connected to a single neuron of an anaesthetised animal can record its sensitivity to frequency, temporal patterns, sound thresholds, and sound pressures. By recording from nerve fibres, both from the ear and in the central nervous system, an understanding may be gained of how complex sounds are being processed and recognised in the brain.

One way in which sounds might be recognised is to specialise the ear itself, right at the beginning of the reception system. It would only be stimulated to respond to a call carrying the correct information. This might be called peripheral specialisation. Another way is to have an ear that picks up everything and then leaves it to the brain to sort it all out. Humans, for example, have an auditory system sensitive to a broad band of frequencies. Our ear does not show any specific selectivity for particular kinds of sounds – we are sensitive to music, speech and the calls of many other animals. Selection takes place in the brain. Specialisation at the 'ear-end' means the brain has a simpler job, but at the cost of limitations in the kinds of sounds that can be heard.

Bob Capranica has found that the frog's auditory system is specialised at the ear. The frog ear is tuned to detect, say, the mating calls of its own species, and not the mating calls or any other calls of other species. The peripheral specialisation is species specific, so the ear of a bullfrog is tuned to different frequencies than that of a tree frog. In addition, in the cricket frogs, the ear of the female is tuned selectively to the local dialect. A female cricket frog in central Texas is completely deaf to the calls of males from east or west Texas.

A further degree of selectivity in the ear has been found in another tree frog from Puerto Rico, called *Eleutherodactylus coqui.* It has the specific name 'coqui' because the male produces a two-note call that sounds like 'ko-kee'. It produces a constant frequency 'ko' note, followed by an upward frequency 'kee' note. Capranica and his colleagues were puzzled to know why a male produces a two-note mating call, as most species studied in temperate regions of Northern Europe and America have only single note calls. Collaborating on this investigation was Peter Narins, now at the University of California at Los

Angeles. It took them to the rain forests of Puerto Rico. There, tree frogs begin to call around dusk, and continue to call throughout the night until the early morning hours. The call, however, is not constant; it changes during the evening. In the early part of the evening, the males set up their calling territories, each male returning to the same spot each night. They are aggressive and will sometimes come to blows if two individuals find themselves too close together. During this period of territory occupation the males produce only the 'ko' note. Neighbouring males will answer each other with this single note. After about an hour of vocal exchanges the males would add 'kee' to their calls, so the forest would echo to the sound of 'ko-kee'.

In order to interpret the two signals Narins and Capranica once more employed playback experiments. A loudspeaker was placed near to a 'ko-kee' calling male and both natural and synthetic calls were played to him. If the entire 'ko-kee' call was played at relatively low intensity the caller paid no attention. He just continued giving his 'ko-kee' call once every three seconds. If the volume of the playback was increased the caller would immediately turn to face the speaker, stop singing the 'kee' note, and respond to the playback with the single 'ko' note. At the same time the caller would approach the loudspeaker. By increasing the volume of the playback the researchers had simulated an approaching male; the louder the sound, the closer the supposed intruder.

Narins and Capranica concluded that there must be some territorial communication going on between neighbouring males. If the 'ko' note alone was played at a sufficient volume to a 'ko-kee' caller he would respond by dropping 'kee' and calling back with 'ko'. If the 'kee' note alone was played, however, the frog would ignore it, no matter what the volume.

The researchers then turned their attention to the female frogs. First, they played back 'ko-kee' calls to which the females immediately responded. They approached the loudspeaker. If the 'kee' note alone was played they would also be attracted towards the sound. If, however, the 'ko' note was played the females paid no attention and often hopped off in a different direction. Clearly 'ko' is a territorial proclamation and 'kee' the mate attraction element. In the early evening the 'ko' note serves to space out territorial rivals. If two males get too close together, they are no longer interested in attracting a mate; instead they become preoccupied in solving the aggressive encounter with the neighbouring male. Most territorial disputes are resolved by calling. The males gradually separate and when the intensity of the call is sufficiently low, each male will resume its 'ko-kee' call and attempt to attract a female. Here, then, is a simple communication system – the male produces two notes; one note signals to males, the other to females. It is possibly the simplest communication system for interaction between the two sexes. And, when the physiology of the auditory system is examined, it is revealed that the ear of the female is selectively tuned to the 'kee' note and the male's ear tuned to the 'ko' note. So, in the auditory system of frogs, Bob Capranica has demonstrated not only species specificity in the sensitivity of the ear, but also geographic and sexual specificity.

Work by Peter Narins and Randy Zelick with *E. coqui* has recently complicated the picture in that female tree frogs have been found to respond to sounds outside the frequency range of the species-specific call. The biological significance of this is not clear, although it could aid the frog in predator recognition – the Puerto Rican screech owl *Otus nudipes* has been found to take *E. coqui* as 30% of its diet. It could also be that the broad auditory sensitivity allows the nocturnal frog to locate its own flying insect prey. *E. coqui* is also quite capable of detecting the first advertisement notes of a related species, *E. partaricensis* which lives in the same vicinity. The researchers propose that our interpretation of species-specific calls of frogs and toads should be modified.

In order to examine the next stage in the hearing process, namely how the information is processed in the brain, Capranica turned his attention to the larger bullfrog. An analysis of its mating call revealed there were three peaks of energy, analogous to human vowel sounds; a low frequency peak around 200 Hz, a trough of low intensity at 500–700 Hz, and a second peak at 1,400 Hz. It was found that, in order for a bullfrog to respond to the mating call, the two peaks of energy must be present. If the low frequency peak is filtered there is no response; likewise with the high energy component. 'How then', asked Capranica, 'are these two energy bands actually recognised by the animal; what is the physiological basis?'

The frog ear differs in one respect from that of birds, mammals or reptiles. In humans, for example, a single auditory organ, the cochlea, converts sound signals into the nerve impulses that go to the brain. In frogs and toads, though, there are two organs in the inner ear; the amphibian papilla and the basilar papilla. The amphibian papilla is tuned to frequencies between 100 Hz and 1,000 Hz, while the basilar papilla is tuned to 1,400–1,500 Hz. For the bullfrog to recognise its mating call, the amphibian papilla must be excited by the low frequency peak in the call at about the same time as the basilar papilla is stimulated by the high frequency component. If either is missing the sound is not heard. The low intensity mid-frequency component mechanically suppresses any excitation of the amphibian papilla at 500 Hz, so this papilla only responds to the 200 Hz peak. Other calls excite different combinations of the two papillas. The distress call, containing only high frequencies, triggers the basilar papilla only. The release call is in the mid-frequency and high-frequency ranges, and so stimulates nerve fibres from both auditory organs.

What then happens to the information when it reaches the brain? Bob Capranica has postulated that the brain of the frog has particular detector sites for each call – a mating call detector, a distress call detector, a release call detector, etc. Dedicated sets of nerve cells in different parts of the brain detect particular sound patterns. Stimulation of those cells by the appropriate combination of frequencies and temporal spacing would somehow lead to an appropriate behavioural response.

Karen Mudry, then working with Capranica, but now at the University of Akron, Ohio, identified two auditory centres in the frog brain – one in the brain stem, in the thalamus, and the other in the telecephalon, which corresponds to the human cerebral cortex. In these two centres Mudray discovered the organisation that had been postulated would be required for a mating call detector. She found an auditory centre that would neither respond to low frequency sounds alone nor to high frequency sounds alone, but which would respond when the two were presented together. Furthermore, the centre would only respond when the temporal features of the sound matched the energy peaks in the mating call. Each bullfrog croak lasts about one second, and is separated from the next by an interval of about seven-tenths of a second. The number of croaks given is variable depending on the maturity of the frog. Typically, a male will give seven or eight croaks and then fall silent for about a minute. It will then repeat the series of croaks. Each croak has a gradual onset time in that it takes about one-tenth of a second for the croak to reach maximum intensity, eight-tenths of a second of constant intensity, and then a gradual decay lasting one-tenth of a second. In the auditory centre found by Mudray, the acoustic envelope required to stimulate that centre must have a rise time of 100 milliseconds, as in the natural sound. The work continues to track down a similar detector in the frog brain for distress and release calls. But, could this amphibian model have wider implications? Is this the way a brain recognises a complex sound? Or is the simple notion of a detector in the brain for sounds naive?

Female Choice

'Explosive breeders', such as the European common toad *Bufo bufo* and the common frog *Rana temporaria* do not seem to 'choose' mates. Males and females arrive simultaneously, in large numbers, at ponds whereupon the males grasp any moving object and hope it is female. The common toad does not have a mating call as such, relying on the release call for sex recognition. Females have very little opportunity to select individual males because they are often grabbed by a searching male *during* migration to the pond. Male European common frogs do have a clearly audible call but playback experiments have not shown that this is effective in attracting a female. More likely the call is to excite the female and encourage her to spawn. Many paired males have been heard to call as well as single ones. In the African clawed toad *Xenopus laevis*, the playing of recorded male calls alone to single females is enough to cause spawning.

In many other species of frogs and toads, females do the choosing, and recently researchers have become interested in the way a female discriminates between individual males of her own species. Some males, it seems, make better mates than others. Females might prefer to mate with older males because age itself is a good indicator of a male's ability to survive over many years, and therefore an indicator of good quality genes. It is becoming apparent that females are able to select mature males simply by listening to their voices. When a female enters a pond of calling males during the mating season she swims around, apparently randomly, seemingly listening to the calls of the males. After a while she will select one individual, head towards him, and eventually spawning will take place. What features are in that male's voice which are preferred over those of the calls of all the other males? Why does the female head for that caller, rather than a neighbour? Could it be the pitch of the call which is important?

In many species of frogs and toads, as the male matures and becomes older he also becomes larger. As he gets larger, so too does his vocal apparatus. Increase in the size of the vocal cavities gives rise to louder calls, while increase in the size of the vocal chords results in deeper pitched croaks due to a lower frequency of vibration of these structures. In humans, the same happens. As we go through adolescence the pitch of the male's voice drops. It appears that in some species of frogs and toads females are more attracted by a call of lower pitch than one of higher pitch.

Michael Ryan, now at Cornell University, New York, sat beside a pond on Barro Colorado Island, Panama, for some 180 nights. He was keeping track of which male tungaras or mud-puddle frogs *Physalaemus pustulosus* were most successful in attracting females. He found that big males get more matings than smaller males. Laboratory playback experiments showed that, of eight females presented with a choice of high pitched and low pitched calls, all eight selected the low pitched sounds.

In similar experiments with North American tree frogs, Carl Gerhardt, of the University of Missouri, found that it was not the largest tree frog which attracted the females, but 'Mr Average'. When presented with a straight choice of two different sounds from two loudspeakers the females were definite in their choice. However, if provided with four loudspeakers, a slightly more complex situation similar to that encountered in the wild, the female's performance deteriorated. This led Gerhardt to suggest that, although the female's auditory system is capable of making fine discriminations in ideal situations, her ability to realise this potential in the wild might be quite a bit lower than had been found in the laboratory. Gerhardt went on to postulate that female green tree frogs prefer 'average-frequency-calls' because of the presence of two other species of tree frogs which have mating calls above and below the frequency range of green tree frogs. The barking tree frog *Hyla gratiosa* calls with a slightly lower frequency, and the squirell tree frog *H. squirell* has a slightly higher pitched call. If female green tree frogs were to prefer a particularly high or low call from the male it might accept a mate of the wrong species, producing hybrids of low viability. There are constraints, it seems, on how 'choosy' females can afford to be in nature.

Anthony Arak, of Cambridge University, carried out similar frequency discrimination tests with European natterjack toads *Bufo calamita*, an amphibian on the British list of endangered species, and one of the few in the British Isles that makes any sound at all. Female natterjacks were exposed to playbacks of high and low pitched sounds and showed no preference at all. Arak then gave them a choice of quiet and loud calls, and generally the females chose the louder calls. This was an interesting result for loudness, too, can be correlated with maturity and body size. Arak also played back the calls at different repetition rates and found that females preferred calls at faster rates, although under normal circumstances they are unlikely to encounter such variations. In the wild, louder or faster calls might not actually attract the females. Laboratory experiment results may simply represent a super-stimulus effect where, if a stimulus is produced at a higher level it is preferred to one at a lower level. Anthony Arak went on to see if the laboratory observations of female choice coincided with those made in the wild. He looked at which males in the population obtained the most females. Large male natterjack toads in a population were seen to obtain as many as eight females during the breeding season. Much smaller males obtained fewer females. Arak tried to find out why. He measured different aspects of the male's calls, whether it was present at the pond on some nights and not others and so on. He soon found that male natterjacks who turned up at the pond on more nights were most successful, although these males were not larger than average. However, he also found that, on any given night, larger males were more likely to mate because they called more loudly and more frequently than the smaller ones.

The larger males, though, do not have everything their own way. Some smaller individuals *do* gain access to the females, and they have gone about it in a rather clever way; it is called the satellite or 'sneaky' male strategy. What happens is that the smaller males wander around the pond until they encounter a large calling male that is likely to attract the females. The smaller frog sits in its immediate vicinity and remains silent. If the little frog called it would be chased away by the big one. When the female approaches the resident large male, the sneaky satellite male nips out and intercepts her, grabbing her in amplexus before she reaches the larger male. In ponds, often only one-third of the male frogs call, the rest remain silent. Sometimes the large frog spots the satellite male as it reaches the female and a fight ensues. A satellite male, though, once firmly grasping the female is difficult to dislodge. Females seem to avoid satellite males, if at all possible. Female tree frogs sometimes shake the little sneaky males off. Female bullfrogs duck underwater and swim below the satellite male, surfacing nearer to the territorial caller. Any toad not calling that approaches a female natterjack toad will cause it to swim rapidly towards the centre of the pond.

Females may avoid satellite males for a variety of reasons. Maybe they are genetically inferior. There is also the danger of mating with the wrong species. Since a satellite male is silent the female does not know if he is a male of her own species. Sometimes several species breed alongside each other, for example, common toads and natterjack toads. Female natterjacks must beware of the promiscuous 'explosive-breeding' male common toads, which grab anything that moves. Males do not have to be so fussy. A male toad has little to lose through a hybrid mating because he can always try again. The female would have spent a year feeding in order to mature her eggs. A hybrid mating would result in non-viable spawn and her efforts would be wasted.

Silent males are not silent all the time. If a calling male is removed the neighbouring satellite male may choose to start up. It is also possible to change a calling male into a silent satellite male by playing his call back at a louder level; when outshouted it seems best to shut-up.

In the case of tree frogs, calling perches are at a premium. The best places are aggressively defended. If kicked off or unable to obtain a perch, a satellite relationship with the frog in residence is the next best strategy to adopt. Satellite tree frogs studied by Carl Gerhardt in North America were equally successful to calling males at obtaining females.

Satellite males, in the main, tend to be young and small creatures, incapable of producing loud calls, and therefore obviously less attractive. If a female is going to discriminate against such a caller it is best for him not to utter a word until he reaches maturity. The satellite strategy, although less productive in some species (satellite natterjacks do badly, while satellite green tree frogs do as well as callers) is less expensive in energy terms. Calling is restricted almost entirely to the breeding season since it is a serious drain on the animal's resources. Body fat reserves have been shown to be depleted considerably after a heavy fighting and calling period. Calling, though, is energetically less expensive than fighting, but is still about six times as expensive as sitting around or feeding.

In North Carolina one enterprising species of toad appears to have developed another form of 'sneaky' strategy to acquire a mate. Small toads have been reported to have been seen sitting in the cooler parts of ponds in order to lower the pitch of their voices. By positioning themselves there, the young pretenders mimic their deep-calling rivals. Deceptions such as this amongst the Fowler's toads *Rufo woodhousei fowleri* of North America are the first known cases of amphibians using surrounding temperature to make themselves more appealing to potential mates; at least, according to Lincoln Fairchild, of Ohio State University. Other herpetologists have questioned the work. Temperature can change certain properties of a call, including pitch, but in a very insignificant way. It would need to be a large drop in temperature to change pitch significantly. Carl Gerhardt has examined temperature variations on the calls of some species and the largest change so far found is in the order of one to two percent change in frequency per degree Celsius change – a small amount. Frogs, though, are poikilothermic or cold-blooded animals. Their body temperature varies with the surrounding air temperature. In temperate latitudes, therefore, frogs call in the early evening, when the pond is still warm, yet visibility is deteriorating for predators. Most species call for the first few hours after dusk and then, as the night draws in and the air gets colder, they stop calling altogether. In the tropics, where temperatures may not vary, frogs and toads will call throughout the night.

A drop in temperature will slow down the metabolic rate of a frog. It simply becomes more lethargic, which means it is unable to call as quickly or as effectively. The pitch and temporal pattern will vary. Female frogs and toads have overcome this problem by locking onto the male's temperature frequency response. As the temperature fluctuates, so the preference goes in favour of higher or lower pitched calls, whichever is appropriate. No matter what the temperature the female always knows at what frequency a male of her particular species should be calling.

In the Canary Islands, the male Mediterranean tree frog *Hyla meridionalis* gives a call which is temperature dependent and the females respond preferentially to certain temperatures. Hans Schneider, at Bonn Universität, has been analysing these calls and has found that the optimum temperature at which the frogs do their calling is between 12° and 24°C. At 24°C a call is brief, about 300 milliseconds, and the interval between calls is just one second. Each call consists of 37 impulses of sound produced at a rate of 125 impulses per second. As the temperature drops, as night falls for instance, the rate changes. At 12°C the call lasts about half-a-second and the interval between calls extends to 3 seconds. Each 12° call has 43 impulses which are delivered at the slower rate of 90 impulses per second.

In playback experiments with artificial calls, in which the parameters of the calls could be altered, Schneider found that females did not prefer callers with the same body temperature, but would move towards callers with a slightly higher temperature. As the outside evening temperature falls, the males sitting in the water lose less heat (and indeed, produce excess heat during calling), than the females approaching the pond overland. A female, suggests Schneider, moves preferentially to a male which, through its call, is indicating that it is at a slightly higher temperature than the female and is therefore probably in the right environment, i.e. the pond, for mating to take place successfully.

Male Competition

The common toad *Bufo bufo* in Britain is unlike many other frogs and toads in that it does not often use a call to attract females. There is some evidence that in this and a few other species the mating call has been lost in the course of evolution. It has been suggested that because common toads always go back to the same ponds, year after year, even to the extent that there are records of toads returning to a place where the pond has long since vanished, they don't need calls to tell them where they should aggregate to breed. If they are all heading for the same place, they will meet up anyway. In early spring huge numbers of toads can be found migrating to their breeding ponds. They invariably take the same route, and at major roads 'toad-crossing' patrols now ensure their safe journey. Whether the common toad has actually lost its call during evolutionary history, or whether it never had one, is not clear. Some toads do make a soft croak in addition to the release call, but there have been no studies to establish whether this is a mating call.

There *have* been studies on another call, a soft peeping sound, which males use in the presence of other males and which functions as a special form of release call. The studies were made by Tim Halliday, from the Open University, Milton Keynes, who was interested in working out which toads are successful in reproduction. In a pond of toads, males outnumber females by about three to one, and Halliday wanted to see whether large or small male toads had females, and which size kept its prize. He found that the larger males tended to hang on to their mates while the smaller ones were often displaced. This led Halliday to examine the way these toads fought for dominance.

A female approaching the pond is intercepted by a male. He leaps onto her back, grasps her behind her front legs with his, and holds on. Any other male interested in taking his place will need to force him from his position. A large male, after a prolonged fight, is likely to dislodge a smaller toad. A fight might last for six hours or more. During the fight the males produce a soft peeping sound, but only when there is actual physical contact between them. Halliday wondered whether these calls were somehow involved in the fight procedure; perhaps the call is saying something about each of the males, maybe about their relative sizes. If a large male can indicate to a smaller one that it is more powerful and inevitably will win a fight, it may induce the small male to move aside and stop fighting. Halliday and his colleagues recorded the calls from a variety of males of different sizes and noticed that there was a direct relationship between the pitch of the call and the size of the toad. Large toads had low pitched calls and vice versa. Did they use this information in fights? To test for this the researchers muted a medium-sized male toad on a female's back by passing a rubber band across its mouth and under its armpits, like a horse's bit, so that it couldn't produce calls of its own. Recorded calls were then played to the pair in amplexus. The researchers then waited for a second male to attempt to dislodge the muted male. When the second male attacked the recorded sounds were played. The attack was much more intense if the high pitched calls of a small male were played than for deeper pitched calls. The attacker appeared to be responding, not to the actual size of the muted male, but to its apparent size on the basis of the call. That, though, wasn't the complete story, for the physical strength of the defender was apparently being registered by the attacker. The attacker was receiving a collection of cues, of which the call was but one.

Many tree frogs give encounter calls which seem to act as a kind of threat. If a defending male on a prime singing perch on a bush or shrub is approached by another male, it will cease producing a mating call and begin an encounter call, a pulsed trill. If the aggressor does not pull out, a wrestling match ensues. Eventually one frog will fall off the perch and the other, the winner of the bout, begins to call again.

Anthony Arak has been studying the tree frogs in Sri Lanka. In one species *Philautus leucorhinus*, the male defends a perch on prominent blades of tall grass. When two males get close they give a pulsed threat call and then proceed to match each other's calls in an escalating singing contest which may last for ten minutes or more. If neither male backs down, the event ends in a fight. Arak could find no pattern related to male body size that correlated with the way residents were or were not supplanted. He compares two call-matching male tree frogs to two bidders at an auction. They progressively increase the number of notes in their calls in response to a rival male's last bid. It is not clear, as yet, whether they are bidding for females sitting near the calling perches or simply bidding to secure their perches.

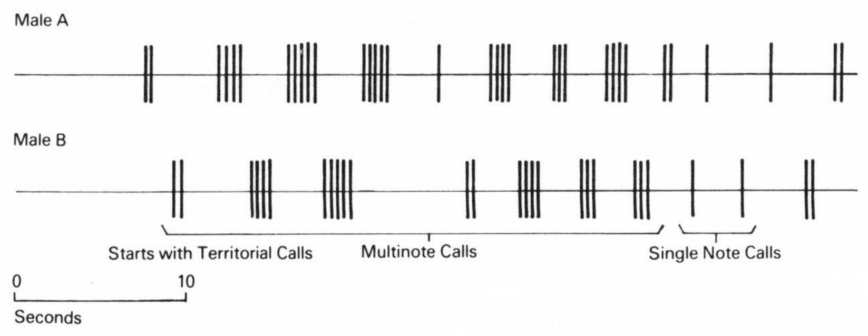

Fig. 33 Male tree frog vocal interactions

Male tungara frogs do not have a territory in the classic sense, but they, too, interact with each other. They use a call which sounds a little like 'ee-y-ur', and get so upset that they turn and face away from each other. These miniature frogs, the size and shape of a square ping-pong ball with tiny legs sticking out at the four corners, back towards each other, kicking fiercely with their hind legs, until one gives up and goes away.

Another group of tree frogs from Panama have complicated two-part calls similar to those of the tungara frog; introductory whistles are followed by a series of click-like notes. *Hyla ebraccata*, for example, has a long introductory note, 'eeeeer' followed by maybe two, sometimes four, clicks: 'eeeeer-i-i'. Males calling by themselves give only the introductory note. Only in competition with other males is the staccato series of clicks produced. Essentially it is a competitive signal. What one male emits determines what another produces.

Kentwood Wells, of the University of Connecticut, has been studying these and other *Hyla* species of tree frog. Originally he thought that these frogs may be matching calls, much like Arak's Sri Lankan tree frogs. If a male gives two clicks then the neighbour may reply with two. There could be an escalation of clicks, and a female arriving on the scene might compare the would-be suitors and pick the frog with the highest number of clicks. Wells believes now that this is not what is going on. Further study has revealed that frogs overlap their calls with those of their neighbours. When a male hears his neighbour begin to call, he waits for a set period and then starts to sing his own song. The neighbour's introductory notes are clearly heard but his terminal clicks are masked by the second male's introductory notes. This is an interference strategy. If the clicks are attractive to females the second male makes it difficult for the female to home in on his rival by masking the attractive part of his call. In addition, the second male may give enough click notes for his call to continue well after the rival has finished. Maybe the last male frog to be heard gets the female.

Many researchers have speculated that, in frog choruses, there is a dominant male who starts things off, and that this male is the most attractive frog in the group. Whitney and Krebs studied calling in the Pacific tree frog *Hyla regilla.* These little frogs congregate at breeding ponds and sing in choruses. There are calling periods alternating with periods of silence. Each bout of calling, though, is initiated by a bout leader, which is not only the first to sing, but also the last to stop. A bout leader may call faster than the rest of the chorus and is more likely to break out into a solo between choruses. In one experiment, a female was placed in an arena between four loudspeakers, each playing back identical calls. One loudspeaker was made to start and stop the chorus of four. In 14 out of 18 playbacks the female headed towards the 'leader' loudspeaker. Whitney and Krebs had shown that the frog which starts and finishes is the most attractive to females. Kentwood Wells' argument goes even further. Simply on acoustic grounds the last male to call may be the most perceptible to the female and therefore attracts the female for successful mating. Wells emphasises that this is only an hypothesis, yet to be tested.

In a tropical forest the sounds created by frogs can be deafening. A person would have to shout loudly to be heard above the din. As often as not, there are many species of frog calling together. The problems each species has in hearing its own calls are immediately obvious. For example, calling in frog choruses with *H. ebraccata* is usually *H. microcephala* and *H. phlebodes*. Unfortunately all these tree frogs broadcast in the same frequency band. It is as if several radio stations were all playing on the same frequency and interfering with each other. All three tree frogs can hear components of each other's calls. If an individual male calls for a female how is she to pick out its call from all the others? Wells discovered that *H. microcephala* is an extremely loud tree frog. A crescendo of these frogs will cause *H. ebraccata* to stop calling. Now, the *H. microcephala* call lasts for about 20 seconds, so *H. ebraccata* males wait for the periods of silence between 20 second calls and sing their songs in the gap.

Wells has demonstrated that all these species of tree frogs are not sitting in the forest in isolation but they hear each other, respond to each other's calls, and this determines how they will communicate with others of the same species. Most work until now has been concerned with the way calls which differ in temporal pattern or frequency structure serve as an isolation mechanism to ensure that a female always responds to the call of her own species. The ways in which a species overlaps or interferes with another have yet to be studied. Given that many frog choruses in the tropics have maybe ten or 15 species of frogs calling, interspecific interference is likely to be widespread. Wells feels that research might usefully turn its attention to similarities in calls, for instance, rather than differences.

The Story of the Frog and the Bat

A male frog's call may not only attract females or repel rival males, but will also draw the attention of predators. Snapping turtles, for instance, are known to home-in on the calls of bullfrogs. For the mud-puddle frog on Barro Colorado Island, danger swoops in from the air. The aerial predator is not a hawk or heron, but a frog-eating bat.

The mud-puddle frog of the rain forests is a miniature animal about an inch long. It has an attractive mating call which, unlike those of many other frogs and toads, is not just a monotonous call; it has a variety from which to choose. A single male, by itself, goes 'aow, aow' and so on, making up to 7,000 calls in one evening between sunset and two a.m. It is using the call to attract a mate. If, however, there are several males within earshot – there may be ten males in a pond three feet long by a foot across – then the caller produces the sound 'aow-chuck' or 'aow-chuck-chuck' and so on up to as many as five or six 'chucks' at the end of the line.

Stanley Rand of the Smithsonian Institute, Washington DC, was intrigued to know the functions of the elements of the two types of call. He made playback tapes – one with the 'aow' call and another with 'aow-chuck-chuck'. Females were then released into an arena and the two calls were played. If 'aow' was played, the female would move towards the loudspeaker. She would also approach if 'aow-chuck-chuck' was played. Rand edited out the 'aow' segment leaving the 'chuck-chuck'. The female was not interested. If, however, she was given a choice of 'aow' and 'aow-chuck-chuck', she would choose the latter. But why, thought Rand, did a male which had gone to the trouble of calling 7,000 times in an evening ever give anything that was less than maximally attractive?

The 'chuck' part of the call, it turns out, contains a broad band of frequencies, is hard-edged, and therefore easily locatable. By adding the 'chucks' the male would simply make himself easier to find; or so it seemed.

Also working at Barro Colorado was Michael Ryan. It was he who noticed that females prefer male frogs with deeper voices. In further tests he showed that females could, in reality, locate a male's call without 'chucks' just as well as a call with 'chucks'. He reasoned then that the 'chucks' contain the preferred drop in pitch which goes with larger males. Ryan carried out more

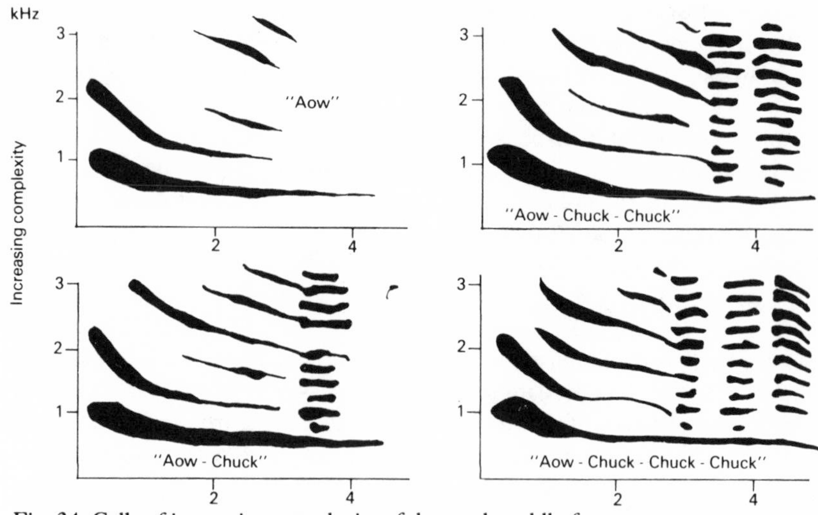

Fig. 34 Calls of increasing complexity of the mud-puddle frog.

tests. He prepared a series of recordings where the pitch of the 'aow' remained constant and the pitch of the 'chuck' was varied. Given a choice of high or low pitched 'chucks', the female chose the deeper sound and thus the larger frog. So the females are, indeed, attracted to a male providing size information in its call, i.e. one giving deep-pitched 'chucks'. The disadvantage, though, of adding this locatable cue is that it makes it easier for a predator to locate the caller too, and it was not until bat researcher Merlin Tuttle, of the Milwaukee Public Museum, began studies on Barro Colorado Island that a predator was identified. One day he caught a bat with a frog in its mouth. It was the fringe-lipped bat *Trachops cirrhosus.* Merlin Tuttle and Michael Ryan went on to unravel the story of the frog and the bat.

The first question Tuttle and Ryan addressed was, might predation play some role in the evolution of the complex vocal repertoire of mud-puddle frogs? In particular, do males leave out 'chucks' at the ends of their calls some of the time because of the danger of attracting bats? Their first experiment showed that the bats do respond to the sounds of the frogs. Tuttle and Ryan went out into the forest and, using a night-vision-scope, watched the frogs at their breeding ponds. They could see that the bats fed quite frequently on the calling frogs, swooping down and pulling the frogs out of the water. They also noticed that the hunting success of the bats was much higher when the frogs were calling than when they were silent. Back in the laboratory, Ryan carried out a series of playback experiments using a large flightcage. A bat was captured and then released into the cage. Frog calls were played through a loudspeaker. By using the speaker, the researchers had eliminated the shape of the frog as a visual cue for the bat and had also ruled out the use of echolocation. When the frog calls were played the bat flew directly towards the speaker, often ripping at the covering in an attempt to get at the frog inside. They repeated the experiment in the forest. Through the night-vision-scope they could clearly see bats heading for the loudspeaker, attracted by the calls of the frogs.

Bats, however, have an auditory system adapted to ultrasonic frequencies; the middle ear apparatus, for example, has bones reduced in mass so that they vibrate more easily at very high frequencies. How can a bat detect the low frequencies of frogs? Tuttle and Ryan played pure tones, over a variety of frequencies, to fringe-lipped bats and found that they were sensitive to the frequencies at which the frogs were calling. Although this species of bat has an auditory system sensitive to its own echolocation calls, it is also capable of hearing lower frequency sounds. The same may well apply to other species of bats.

But there are many kinds of frogs and toads living in and around the ponds of Panama; some potential food for the bats, others poisonous. Could the bats pick out one frog from another, simply by their calls? In further playback experiments, bats were given a choice of calls. One speaker played the calls of a species of frog known to be predated by bats, while the other played the calls of toads of the genus *Bufo*, all of which have poisonous skin secretions. The bats preferred to attack the speaker playing the call of the edible frog, and for the most part ignored the speaker with the call of the poisonous toad. A similar experiment was carried out with the calls of the South American bullfrog *Leptodactylus pentadactylus.* Some individuals of this species grow to a foot in length from snout to vent and are much too large for the bats to capture. Occasionally these giant bullfrogs will turn the tables on bats by leaping out of the water and capturing them as they fly by. Ryan gave bats the choice of the calls of bullfrogs and their preferred edible frogs. They avoided the bullfrogs. Evidently, bats are able to distinguish one species of frog from another purely on the basis of the call.

If the frogs, then, are to take evasive action they must be able to find out when an attack is coming. If bats can detect frogs, can frogs detect bats? It seems that they can. Tuttle and Ryan recorded a large chorus of frogs and noted what happened to the calling behaviour before and after a bat flew over. If it was a moonlit night or a night with no moon but no cloud cover, the frogs would stop singing immediately a bat flew across the pond. On a totally dark moonless, cloudy night, the frogs showed no response to a passing bat and continued calling. To show that the frogs were relying on visual cues, a model of a bat was flown over an experimental pond of calling frogs. By controlling light intensity, simulating light and very dark nights, it was demonstrated that the frogs ceased calling if they spotted the model flying over, yet ignored it if it couldn't be seen. The researchers also eliminated the bat's echolocation calls as a warning cue for the frogs. High frequency sounds, similar to those emitted by the bat, produced no response from the frogs.

Two species of frogs on Barro Colorado Island form the main part of the diet of fringe-lipped bats; one is the mud-puddle frog and the other the pug-nosed tree frog *Smilisea sila*. Tuttle and Ryan have found that there are two quite different, almost opposite, strategies to avoid capture by bats. The mud-puddle frogs gather in large groups and sing in chorus. If there is a constant level of predation, the chances of being eaten are relatively low if an individual is surrounded by many others—the 'selfish-herd' effect. The pug-nosed tree frog, on the other hand, does not call in choruses but is spaced out every five metres or so along stream banks. It, too, can spot the approaching bat, but behaves quite differently according to whether the night is bright or dark. On bright, moonlit nights the frogs sit and call out in the open, on the tops of rocks, and remain there for much of the night. On nights without a moon, when they are unable to see the bats flying down the stream, they start to call at dusk but then quickly move to the tops of trees, or stop calling altogether and hide under rocks, fallen tree trunks or large leaves. They decrease their exposure to the bats when they cannot detect them.

10

SEEING BY SOUND

Bats fly at night, and for centuries have been the objects of fear and superstition. These nocturnal wanderers were clearly in collusion with the powers of darkness, evil devils to be found in the company of witches and warlocks. They got in your hair and were an essential ingredient of witches' brews. They sucked the blood of sleeping men. But their ability to fly in the dark has also fascinated generations of naturalists.

One of the first of these was the eminent 18th Century Italian naturalist, Lazzaro Spallanzani. In a series of important papers on his discoveries in different fields of natural science, he published some remarkable observations on bats. In the Autumn of 1793, Spallanzani discovered that bats are able to detect and avoid obstacles in their flight path even when they cannot see. If the translated reports of his experiments, by Robert Galambos and Sven Dijkgraaf, are anything to go by, Spallanzani's research was extremely cruel and would in no way be condoned today. He did, however, show amazing insight and understanding of the baffling results he obtained.

The first written accounts of the bat experiments, in 1794, were in an exchange of letters between Spallanzani and Abbé Vassalli in Turin, Pietro Rossi in Pisa and Jean Senebier in Geneva. He encouraged them to repeat his experiments. Bats were brought indoors and their eyes destroyed, either by putting a hot wire through the cornea or by pulling the eyeball out and cutting it away from muscular attachments. Vasalli poured hot wax over the eyes which then cooled and set. To make quite sure, Rossi placed opaque material in the socket and pasted it over with a leather disc. Animals avoided obstacles equally well before and after the mutilation. Spallanzani wrote:

> 'Sometimes the animal, suffering from the operation, flies with great difficulty; later it can be made to fly freely in a closed room either during the day or at night. During such flight, we observe furthermore that before arriving at the opposite wall, the bat turns and flies back dexterously avoiding obstacles such as walls, a pole set across its path, the ceiling, the people in the room, and whatever other bodies may have

> been placed about in an effort to embarrass him. In short, he shows himself just as clever and expert in his movements in the air as a bat possessing its eyes'.

Touch was the first explanation put forward. The tips or edges of the wing might brush an obstacle, whereby the bat would quickly veer away; or, air currents caused by the flap of the wings might bounce back on hitting an object. Spallanzani quickly rejected touch. Bats would fly in curved tunnels without touching the walls and could pass between silk threads. Varnish of sandarach and spirit of wine painted on the wings confirmed that touch was not involved.

Taste and smell were rejected by Spallanzani in his usual forthright way – he cut out the tongue and plugged the nostrils. He and Vassalli did notice, however, that there were complications if the nose was plugged, but it was put down to an impairment of breathing.

The head region, though, appealed to most of the group as the centre of activity. An opaque hood was put over the head of the bat and the animal was unable to fly. Hearing loss was not suspected and an unknown and mysterious sixth sense was proposed as the sensory mechanism involved, although Spallanzani was never really satisfied with this theory.

Louis Jurine of Geneva, in the meantime, had repeated Spallanzani's experiments and concluded that a bat's ability to avoid obstacles was intimately linked with its capacity to hear. Jurine plugged the bats' ears with turpentine, wax, pomatum or tinder mixed with water and they were unable to fly properly. Jurine concluded:

> 'First, that the eyes of the bat are not indispensably necessary to it for finding its way; secondly, that the organ of hearing appears to supply that of sight in the discovery of bodies, and to furnish these animals with different sensations to direct their flight, and enable them to avoid those obstacles which may present themselves'.

Spallanzani concentrated on hearing, and was fully convinced, until the time of his death in 1799, that bats orientate in flight by using their ears. How it was achieved he was unable to discover.

In 1906, A. Whitaker noted that 'bats are not disturbed by man's speech, but are greatly disturbed if the hands are clapped together or paper is torn'. At that time, it was thought that bats might be picking up echoes from sounds associated with their wing beats, and using these as a method of finding out what was in front of them. The idea was rejected, eventually, because the sounds involved with bat wing beats are very low in frequency, so low that the bat could not gather enough information from them.

In 1920, the English physiologist, Hamilton Hartridge, suggested that bats might get around in the dark by using high frequency sound. He said little more than that for it was simply an idea; he didn't have the technological back-up to take it further. But in the late 1930s, Professor Don Griffin, now at Rockefeller University, New York, took an interest in bats and this was to lead to the revelation that bats, indeed, could see with sound. Griffin was interested in the migration of bats and, inevitably, was confronted with the fact that bats are able to navigate and orientate in totally dark caves. He was at Harvard at the time and became acquainted there with the physicist Professor G. W. Pierce, a pioneer in the electronics field. (Among other things Pierce discovered the principle of using a crystal to stabilise a radio-frequency oscillator.) Late in life he developed an interest in the sounds made by insects, most of which are in the high frequencies, some even above the human hearing range. For the purpose of listening to very high-pitched insect sounds, Pierce constructed a special piece of apparatus that would detect sounds above the frequency range of human hearing. Griffin saw the potential of Pierce's apparatus and took a cage of bats to be tested. Sure enough, the apparatus picked up a variety of sounds. In 1940, Pierce and Griffin reported that bats emit ultrasonic, high frequency sounds when active. Next they encouraged the bats to fly. This was difficult for if a room was not specially prepared the bats would find somewhere to hide, and ruin the experiments.

The first flights were in Pierce's sound proof chamber, a room about eight feet by 16 feet, but the results were very disappointing. When the bats were flying, Pierce and Griffin couldn't detect the same sounds that they had recorded from bats in the hand, which were held right in front of the microphone. Their first paper, then, was rather guarded. Bats, they said, were capable of producing 'supersonic' calls, as they were termed at that time, but it was not clear if or how they were used in flying. Griffin then met a fellow student, Robert Galambos, a physiologist working on the neurophysiology of hearing with Professor Davis of the Harvard Medical School. Together, Griffin and Galambos began to work on bat acoustics. Pierce allowed them to borrow his apparatus and it quickly became apparent that they had made an unfortunate mistake in the way they were attempting to listen to the flying bats. The microphone they were using was mounted in a deep, parabolic horn which made it very directional. What they hadn't realised was that the bat's sonar is also highly directional. The microphone would have only picked up the bat sounds when the creature was flying directly at the horn. By moving the apparatus around Griffin and Galambos discovered their mistake, for when the microphone was dead ahead the bat came in loud and clear. They had discovered that a bat's ultrasonic calls are used in flight almost continuously. The two researchers continued with a series of classic experiments where bats were flown across dark rooms littered with obstacles. Vertical wires, spaced apart so that the bat could just fly between, were successfully avoided. They then repeated some of Spallanzani's experiments, only a little more humanely. Specially made ear-plugs and masks were used to prevent bats hearing or emitting sounds. If the experimental apparatus registered no sound being produced then the bat couldn't dodge the wires. Likewise, if its ears were plugged it became disorientated and collided with the wires.

In the meantime, Galambos had been investigating the cochlear microphonics of the bat ear. He recorded the electrical signals from the cochlea and found that frequencies as high as 100,000 Hz were being detected. This was a novel finding at that time for nobody realised, as is now well known, that practically all small mammals hear higher frequencies than humans. We, it later transpired, together with elephants and some of our higher ape relatives, are the odd mammals out.

Echolocation

Bats are out at night and at this time their vision is poor. Bats are not blind, as the saying goes, but the conditions under which they operate make vision inappropriate. There are bats, the fruit-bats, Megachiroptera, which use a sophisticated night vision, but even here some light must be available. On a cloudy, moonless tropical night fruit-bats do not fly. The small, mainly insect-catching bats, Microchiroptera, are night flyers, and they can fly in the dark. To do this bats use echolocation. High frequency sounds or ultrasounds emitted by the bat are bounced off targets. The returning echo is received and analysed by the bat which gains information about the location or direction of movement of the target. Ultrasound has characteristics that make it suitable for such a task. The wavelength is short so that it can be projected from a small transmitter, and it will provide more directional information than a long wave. Short waves are also more easily reflected from small objects.

There are two ways in which sounds reflected from a target can be analysed – a bat can see how long a sound takes to come back or it can detect the change in frequency in the echo. Both methods are used.

To measure range, for prey detection or navigation, vespertilionid bats emit their high frequency sounds in very short pulses. The sound produced has a wide band width, which makes it more effective for measuring range. Each pulse is frequency modulated, that is it starts on one frequency and rapidly descends in pitch to a lower frequency, covering all the frequencies in between. Very commonly there is a complete frequency sweep, say from 80,000 to 40,000 Hz, and each pulse lasts for only thousandths of a second or milliseconds. No bat in this group has a sound pulse longer than 20 milliseconds or a fiftieth of a second. More usually the duration is as little as one millisecond or less. A cruising bat, in no kind of hurry, may produce something like five to 20 pulses per second, but something Griffin and Galambos noticed early in their research was that whenever a bat had a difficult orientation problem – trying to detect something small or when coming in to land – it speeded up its pulse repetition rate. A bat may speed up from, say, ten pulses per second to 50 pulses per second. This is particularly spectacular when it approaches a target prey insect. A vespertilionid bat may cruise high,

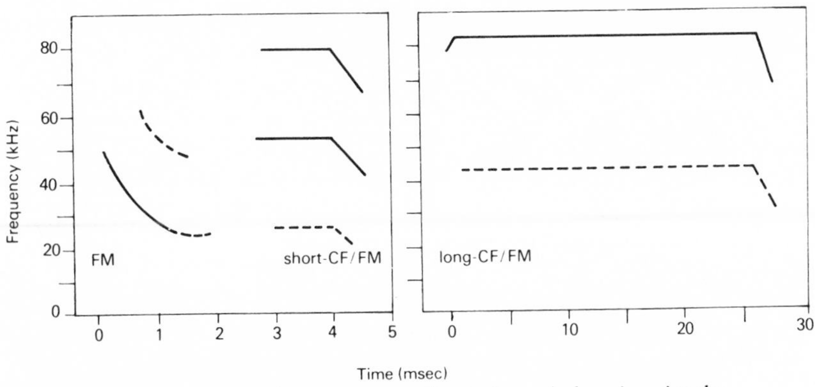

Fig. 35 Diagrammatic sound spectrogram of main bat echolocation signals: frequency modulated (FM), short constant frequency (CF) with terminal FM, and long CF with terminal FM (Doppler). Strong harmonics shown by solid lines.

emitting pulses at five or ten per second, when searching for insects, and then speed up dramatically with 200 per second 'buzzes' as it closes on its target. What the bat is doing is gathering increasingly more detailed information about the size, speed and direction in which the target is travelling. It is more critical for the bat to have this information as it closes in, and its manoeuvring needs to be more precise.

To test that these insect-catching bats are using their echolocation system for range measurement, James Simmons, of the University of Oregon, trained bats to take off from a platform and to fly in a darkened room. In front of the bats were two landing platforms and they were trained to fly to the nearer one. This they did with ease. The landing platforms were then gradually brought together and the bats' ability to measure the small differences in distance from the take-off platform was noted. Moving the targets, though, introduces other factors which complicate the picture somewhat, so Simmons refined the experiment electronically with delay lines. This time the targets were placed at exactly the same distance and were not moved. The echo was delayed electronically. A microphone, near to the bat, picked up the sound and rebroadcast it from a loudspeaker to make an artificially delayed echo. Transmission from microphone to loudspeaker was so rapid that the delay could be made very short. Bats could detect a delay so brief that they were capable of discriminating differences in range of only a couple of centimetres (two centimetres is equivalent to a delay of 120 microseconds). This is an accuracy that can be calculated to be much less than the duration of an individual wave in the bat's sound pulse, and it can be achieved much more easily, both in theory and practice, by using a frequency modulated sweep with a wide range of frequencies in it than with any other signal. Indeed, in a recent paper, Simmons has suggested that the depth of pits in the target platform is detected by analysing the spectrum of the echo.

Further experiments by Simmons and B. D. Lawrence at Oregon have

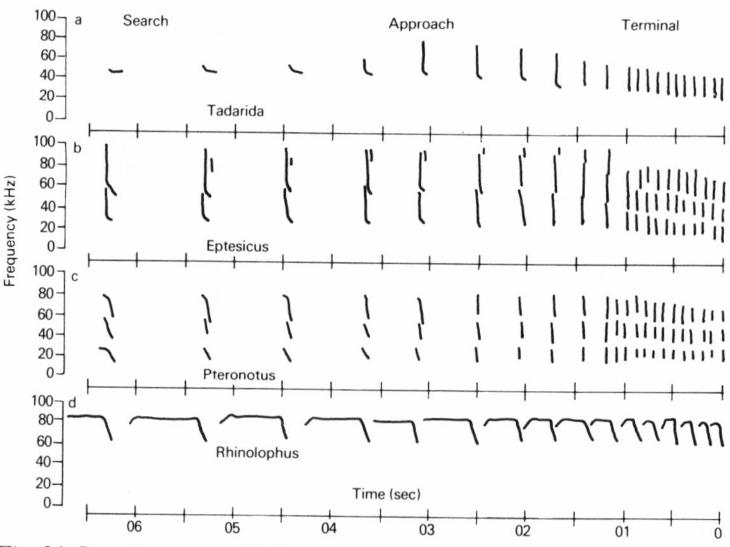

Fig. 36 Sound patterns of different species of bats when approaching a target.

shown that big brown bats *Eptesicus fuscus* can perceive, not only horizontal information, but also vertical cues. An acuity of about three degrees of arc is achieved by modification of the returning sounds by structures in the external ear. The tragus, an inner lobe to the external pinnae, appears to deflect part of the echo, producing a strong secondary echo. The delay of the secondary echo gives the bat information about the vertical direction of the target.

The bats with frequency sweep pulses or frequency modulated sounds have become known as the 'FM' bats.

The second way in which a bat's radar might work has given rise to the name 'Doppler' bats. Motorists who have been caught speeding in restricted areas may have come across Doppler radar. This form of radar operates, not by looking at the time taken for a signal to come back, but by looking at the frequency of the returning signal compared with that of the outgoing signal. A bat would know the frequency it sent out, but the frequency coming back will be a little different if the target is moving with respect to the transmitter or if the transmitter is moving with respect to a fixed target. It is the relative motion that is measured. The round-trip distance is changing continually. If the bat is approaching a target the later sound waves return with less delay than earlier echoes, and so all the waves squeeze up into a smaller space and the frequency goes up. This difference in frequency between transmission and the echo is, therefore, a way of measuring relative velocity between the bat and its target. If the target is moving away the returning frequency goes down; if the target is moving towards the transmitter the returning signal goes up. To achieve accurate ranging with this form of radar, however, we would expect the bat to transmit in as narrow a bandwidth as possible (as bandwidth is not important) and continuously rather than in pulses if possible. But a bat cannot do that. It has to breathe. A bat breathes at ten to 14 breaths per second, which is linked to wing-beat rate in flight. Nevertheless, Doppler bats *do* use relatively long duration monotonal sounds, up to 100 milliseconds, or a tenth of a second per pulse, and can produce these longer pulses for about 90% of the time.

The European greater horseshoe bat *Rhinolophus ferrumequinum* is a 'Doppler' bat. The German zoologist F. P. Möhres found that the greater horseshoe bat emits long-duration, constant-frequency pulses. Drs Ulli Schnitzler and Gerhard Neuweiler, of Marburg University, West Germany, discovered that this bat has a finely tuned auditory system, sensitive mainly to a narrow band of frequencies and that it uses this system to measure the velocity of a target using Doppler radar. The greater horseshoe bat cheats a little by introducing small frequency sweeps at the end of each pulse when approaching a target. Schnitzler, however, showed that the greater horseshoe could detect minute changes in frequency, due to the Doppler effect, caused by the relative motion of the bat and its target. The bat puts out a signal of around 83,000 Hz, and it can detect a frequency shift in the returning echo of just a few hundred Hz. The bat, though, has a narrow frequency window, below its normal emitted frequency, at about 82,000 Hz, to which it is most sensitive. To ensure that the echo arrives back at this window, the bat compensates by changing its emitted frequency.

It was not clear for some time how the hunting tactics of 'Doppler' bats differed from those of their 'FM' cousins but work in 1968 by David Pye of Queen Mary College, London, on East African horseshoe bats and more recently in Australia by Pat Brown, of the University of California at Los Angeles, has revealed one strategy. Some 'Doppler' bats remain at a roost site and hang there scanning the air with a very narrow beam of sound, and rotating the head, much like a ground-to-air missile radar system. When an insect flies into the scanning area the bat flies out, catches it, and returns to the roost. The European horseshoe bats have adopted an additional method of hunting. They skim low over the ground and search for insects that are warming up.

So, there are two extremes of bat echolocation radar – the 'FM' and the 'Doppler' systems – and evolution has refined this behaviour to such an extent that bats exploit these systems to the limits of what is theoretically possible. In between the extremes are intermediate forms of radar, the likes of which are only now slowly being discovered. There may be frequency sweep pulses, for example, but with second, third, fourth or fifth harmonics extending the band width, which sweep simultaneously.

David Pye started, in the summer of 1980, to construct an evolutionary tree based on the echolocation systems used by the various bats, but by the following year it had become clear that bats use systems which suit their needs and that some bats will adapt their behaviour from one extreme to the other according to conditions. One such bat is the European pipistrelle *Pipistrellus*

pipistrellus, an insect-catching bat that has been studied by David Pye. It is the commonest British bat, often seen flying not far above head height on a summer evening. When cruising at five to ten metres, looking for insects, the pipistrelle needs to examine as large a volume of air as possible. It does this using 'Doppler' radar, with a low frequency signal of long duration. This gives the bat good information about the velocity of the target but is no good for telling the distance accurately. The bat, having detected a target, changes over to 'FM' sweeps. This is done gradually at first, introducing a small sweep at the start of each pulse, then eliminating the constant frequency, and deepening the sweep. The strategy ends as the bat closes on its target, with a series of rapid bursts of frequency sweeps which are ideal for measuring exact distances.

Observations such as these can actually be made in the field with the help of a piece of electronic wizardry known as the bat detector. David Pye has developed one such device in order to study wild, free-living bats without any interference apart from the researcher's own presence. The Pye bat detector works by translating the ultrasonic calls of bats into frequencies audible to the human ear. The result is a series of clicks. You can hear what the bat is doing while it is doing it.

Pye is now developing more sophisticated equipment which will record the bat's velocity relative to a recording microphone attached to a small Doppler radar. The frequency emitted by the bat can then be identified accurately regardless of any motion towards or away from the microphone.

The Auditory System of Bats

Many people consider bats to be ugly, even repulsive, creatures, but in reality their unusual appearance is a consequence of their specialisation as aerial predators. The structure which gives many bats their distinctive appearance is the nose-leaf. All bats that produce sounds through the nostrils have, on the face, complicated folds of skin. It is thought that the nose-leaf is focusing the out-going sound. It is, in effect, the transmitting antenna.

The call of the bat is produced in the voice-box or larynx, which differs from the voice-boxes of other mammals in being highly specialised and relatively large for the animal's size. Most mammals have a larynx about twice the diameter of the trachea. Much of this bulk is filled by a greatly enlarged muscle which puts tension on the vibrating structures. Bats have two pairs of very thin membranes, six to eight thousandths of a millimetre wide, which are stretched across a cavity roughly a quarter of a millimetre deep. The cavities are on the wall of the larynx, separated by folds which act as valves. The vibrating structures are like tiny drums with a slit running across the centre of the drum-heads. Air passing across the split membranes cause a sound to be produced. The membranes are attached to the skeletal framework of the larynx which is strengthened with bone. Most other mammals have a larynx with cartilage. The strengthening allows the powerful muscles to stretch the vibrating membranes to a high tension and so produce very high frequency sounds. The lowest recorded bat, so far, emits a basic fundamental tone of about 12,000 Hz, and is just about audible to man. More usually the signal is about 20,000 Hz. One African species was recently recorded producing a sound with a top frequency of about 210,000 Hz. The vocal chords vibrate at about 105,000 times per second, and the high note is produced as a second harmonic, which is accentuated by special resonators in the nose.

The hearing system is, likewise, based on the normal mammalian pattern but with a high degree of specialisation for the reception of ultrasounds. Many bats have large, mobile ears for scooping up the returning sound signals, but not all. Some have small ears. But no matter the size, inside are to be found all the basic components of the mammalian ear with certain modifications. In the middle ear, all mammals have two muscles that pull on the middle ear ossicles – the three small bones that connect the eardrum to the inner ear. The muscles serve primarily to reduce sensitivity. In the bat they are large and seem to function as an on/off switch. This was suggested in 1945 by Hamilton Hartridge but no research was done until recently when William Henson, of the University of Arkansas, discovered that, sure enough, the muscles contract and relax fast enough to reduce sensitivity during, and for a few microseconds after, the production of a pulse of sound. In this way, the bat doesn't deafen itself.

The cochlea turns the sounds into nerve messages which go to the brain for decoding. Despite the small size of the brain in these animals, the auditory centres are relatively large. Those parts responsible for decoding the sound signals in the hind- and mid-brain are enormously developed. The auditory regions of the decision-making fore-brain are not so well developed because the bat is not interested in the quality of the sound; it is more concerned with the picture of the external world that the sound has revealed. The forebrain makes decisions on behaviour based not on the sounds themselves, but on the image decoded in the mid-brain.

Not only is the architecture of the brain highly specialised, but so too are the neural mechanisms within it. There are single brain cells or neurons which will respond to the second of two sounds even though the second is fainter than the first. Humans tend to suppress echoes, unless they are very pronounced like those coming from a cliff-face, that is, bounced off distant objects. Bats on the other hand respond to echoes from very short distances. Neurons in the brain ignore the signal sent and respond only to the fainter echo returning. Sophisticated switching circuits, which parallel the anti-deafening mechanism, ensure that sensitivity is restored as rapidly as possible after transmission so that the brain can appreciate the echo.

Particular cells in some parts of the brain can be shown to respond only to a narrow range of frequencies, coming from a certain direction, and at a fixed time after the outgoing signal was made. One of the leaders in this field of

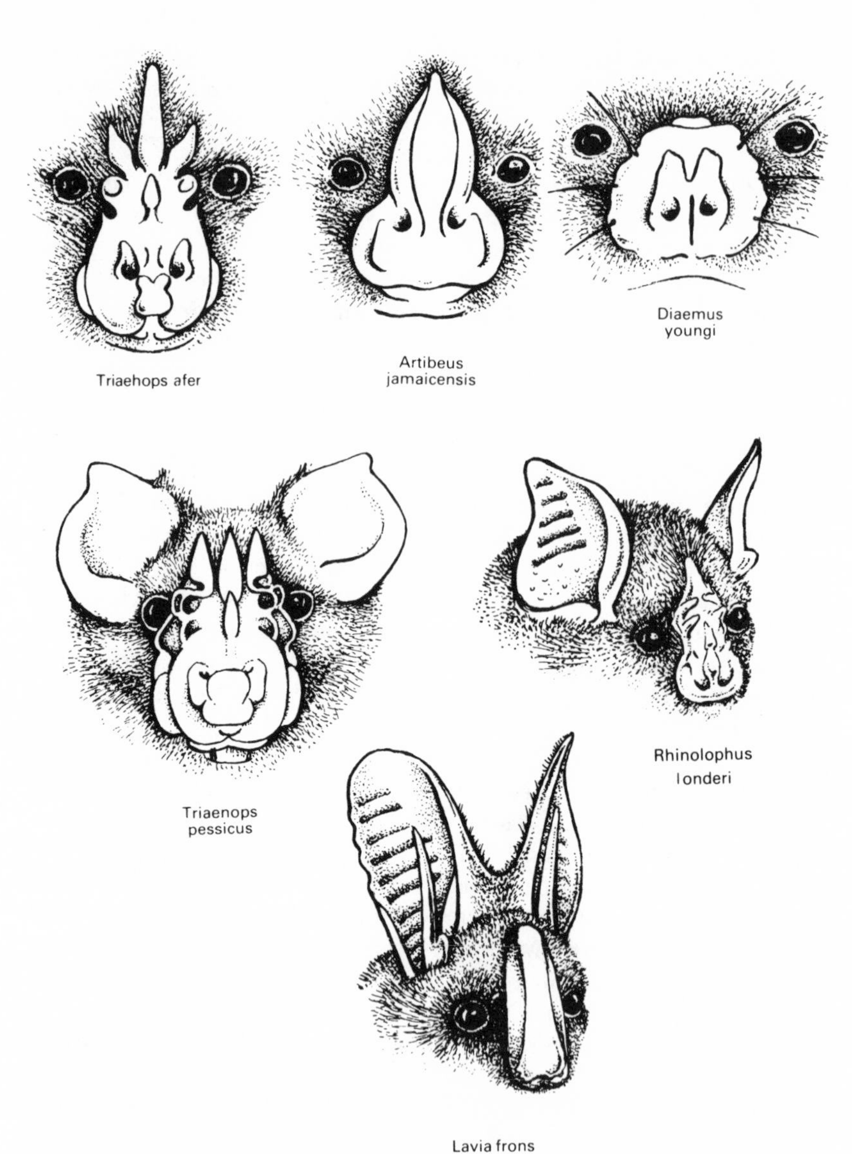

Fig. 37 Heads of different species of bats showing variety of nose-leaves and ears.

research is the Japanese researcher, Nubuo Suga, of Washington University, St Louis. He has been working with a New World bat, *Pteronotus parnellii*. This is an insect-catching bat which, like the greater horseshoe bat, uses constant frequency signals with short terminal sweeps. The frequency is so constant that it requires very accurate instruments to do it justice. At 60,000 Hz the signal is constant within about 50 Hz. In the brain, the cells of the auditory cortex are laid out in a frequency determined geometric pattern, so-called tonotopic representation. A low frequency area is at one side and the intermediate and high frequencies at the other. This is not unusual in mammalian brains, but this bat takes the arrangement several stages further. The dominant constant frequency used by the bat has a large area of the brain devoted to it, and then, radially outwards, other frequencies are represented. Intensity is represented in a similar way, radiating out from faint to very loud. In another part of the brain, Suga found what amounts to a map of the external world in terms of cells that respond preferentially to echoes coming from different directions. In yet another area cells respond according to sounds coming from different distances.

The 'Two-Toned' Bat

For nearly 20 years, David Pye and his colleagues at Queen Mary College have been recording the calls of bats. So far, they have covered over 120 species, representing 16 out of the 19 families. With 850 known species of bats on the planet the researchers have plenty of new material left to work on. Some of the bats that have come under David Pye's scrutiny produce curious signals. One in particular, nicknamed the 'two-toned' bat, demanded further study. There is one genus *Saccopteryx* of two species of New World bats, living in Trinidad and Panama and another not too distantly related species *Coleura afra* in Africa. When hunting, all three appear to produce an almost constant frequency signal with harmonics. If the high-speed tape recording is slowed down considerably, however, the pulses appear to be produced in two quite different and alternating frequencies. As they fly past the microphone, they sound like a passing police car with its two-tone klaxon going. Pye noticed that the difference in frequency between the two pulses was curiously equal to the Doppler shift of echoes returning to the bat from a target straight ahead. So he is now speculating that the bat is operating two separate radar systems on alternate pulses; one looks directly ahead for flying targets, obstacles in its flight path and so on, while the other beams downwards at the ground or sideways at trees and bushes in order to give the bat an orientation with respect to its environment. These bats appear to hunt in thick forest and in this habitat it might be necessary to have two systems. To solve the mystery, the new Doppler radar and recording system, developed by Pye, will be used.

The 'Gleaners'

Although large flying insects are relatively easy for bats to spot and capture, those sitting on walls or leaves are difficult to detect with echolocation. One group of bats specialises in insects that are not flying. They are known as the 'gleaners'. The European long-eared bat *Plecotus auritus* is a 'gleaner' some of the time, but the majority of 'gleaners' are tropical species. In general, 'gleaners' make fainter sounds. Many have large ears. But detecting an object on a flat surface is a difficult operation to carry out with echolocation. A flying bat chasing a flying insect is concerned with locating a hard target on a soft background. The 'gleaner' has the more difficult task of detecting a hard target on a hard background. One team of researchers which has been trying to find out how 'gleaners' glean is from Carleton University, Ottawa. There, Gary Bell, working with Professor Brock Fenton, has been studying pallid bats *Antrozous pallidus* and Mexican long-eared bats *Myotis auriculus*, which occur in the American south-west. They first looked for the neatest solution – a specialised echolocation call. There was none. The answer, as it turned out, is that the North American pallid bats listen for the sounds made by moths and other flying insects as they 'warm up' prior to take-off. Insects, sometimes, have to raise the temperature of the muscles of the thorax before they can fly efficiently, and the pallid bats home in on these wing vibrations. Long-eared bats do, in fact, use faint echolocation calls if it is dark, but given sufficient light they simply use vision. They switch back and forth between the two means of location depending on which is appropriate to the conditions that prevail. They take advantage of any cues available to make themselves more adept at catching prey.

The research team carried out a series of laboratory experiments to find out how bats reacted to conflicting information. They presented the bats with situations where their eyes would tell them one thing, but their echolocation system would indicate another. Connie Gaudet trained pallid bats to fly to a target. She then put up the target but separated it from the bat by a sheet of plexi-glass. The bat flew to the target, smashed into the plexi-glass and returned to its perch. On its second flight it clearly switched on its echolocation system, found its way around the plexi-glass and reached its target. Pallid bats, like their long-eared cousins, are flexible in the way they use their senses and are not exclusively bound to echolocation.

The 'Whispering' Bats

Some small bats do not chase after insects but have alternative sources of food. In the New World, one group of leaf-nosed bats, *Carollia*, have taken to fruit-picking. They have sonar, but of such low intensities that researchers have dubbed them the 'whispering' bats. Their sounds can only be detected if they are no more than a few centimetres away from a microphone. Like the 'gleaners', the 'whisperers' rely less heavily on echolocation and utilise other senses as and when appropriate. The fruit-eaters use vision and smell, for example, to find fruit but still use sonar for obstacle avoidance, particularly if they roost in caves.

Vampire bats are 'whispering' bats. They bite and lick blood from mammals which they locate mainly by smell. Again, they have echolocation but it is faint.

Silent Hunters

Indian false vampire bats *Megaderma lyra* hunt in silence. J. Fiedler in West Germany trained false vampire bats in the laboratory to catch and eat mice. Did the bats use their echolocation calls to pinpoint the mice? Mice have good ultrasonic hearing, and it is conceivable that they could detect the presence of the bat and avoid capture. Fiedler set up a large arena, liberally sprinkled with microphones to pick up any sounds being made, set a mouse loose, and sent in the bat. The bat made a silent approach, swooped, and caught the mouse, and then started to echolocate its way out of the area. Like the barn owl, the Indian false vampire bat detects and catches its prey by homing-in on the noises made by the mouse scuttling through the undergrowth. It switches on the echolocation system simply to clear trees and bushes, and to fly back to its perch where it consumes its meal.

Bats and Moths

Bats have been trying to catch moths and moths have been trying to avoid bats for millions of years. The earliest known fossil bat from Wyoming in the United States was flying about, some 50 million years ago, only 15 million years after the dinosaurs had disappeared. Already, at that time, it had a functional echolocation system. Some researchers, including Brock Fenton, believe that the development of the bat's echolocation system to catch flying insect prey was important in the evolution of both bats and moths.

Many of the juiciest insects, including the large moths, fly at night and echolocation is an excellent way of finding them. Echolocation, however, does have its drawbacks. When we look at modern bats it is interesting to note the intensity of sound being produced. About ten centimetres in front of the little brown bat *Myotis lucifugus*, the sound level is 110 dB, the equivalent of putting your ear next to a very loud alarm bell or siren. It may be emitting these intense, piercing, high frequency sounds at about 20 pulses a second, and when homing-in on a target insect, produces about 200 per second. In effect, the bat is flying along yelling its head off. It is very conspicuous. A conspicuous animal, inevitably, is going to have difficulties catching its prey. Echolocation, then, is not as perfect a system as was first thought. Many insects have ears, and some are sensitive to the echolocation calls of bats. A tiger moth (Arctiidae), for example, can detect the calls of a big brown bat *Eptesicus fuscus* at about 40 metres, but the big brown bat can detect a 19 millimetre, moth-size, diameter target only at five metres. So, the bat ranges on the moth at five metres but the moth has heard the bat coming from 40 metres.

The moth takes evasive action, and it has several ways of doing so. Since the early 19th Century, morphologists have known that moths have an earlike structure on their sides. This is a thin section of the insect's chitinous exoskeleton with an associated nerve cell. To researchers it looked like an ear, worked like an ear, and probably was an ear, but there was no apparent reason for having an ear as moths are not very noisy creatures. Some moths do make sounds but they are not the obvious noises that, say, crickets make. It had been known for some time that if two pieces of metal are squeaked together, so that they produce that nasty high-pitched grating noise, a flying moth, such as a tiger moth, will tumble suddenly out of the sky. People thought that moths were messengers from Satan and that these high-pitched sounds would send them scurrying into the night where demons belong. It was not until Don Griffin discovered ultrasonic echolocation in bats that Kenneth Roeder and Asher Treat of Tufts University, Massachusetts, were able to put two and two together to reveal that moths detect bats by their calls, can find out the direction from which attacking bats are coming, and can take evasive action. Asher Treat's interest was in ectoparasites on moths. Mites get into the ears of moths and damage them. Mite infested moths cannot detect bats. Non-reacting moths were caught about three times as often as reacting moths. Treat instantly recognised the advantage gained by moths without mites. They could hear the bats.

Most moths have the pair of ears on the thorax. Some have them on the abdomen, and yet others have ears on the face. Hawkmoths (Sphingidae) have ears on highly adapted mouthparts. This suggests a polyphyletic origin for the evolution of moth ears, that is, they have evolved many times in different ways in different groups. David Pye suspects that the groups of moths with ears diverged long before the bats appeared, so the different groups of insects had to solve the problem of the new insectivorous bats in their own way. Maybe that is why there is no one place for the moth ear.

The classic evolutionary origin for the moth ear was put forward by Kenneth Roeder. He suggested that it first served as a stretch receptor, a monitor that told the insect, for instance, in which position its wings were, or how much stretch was being put on the exoskeleton. Insects are covered with small sense organs on the inside of the exoskeleton. It would be relatively easy for the exoskeleton to be thinned out around one sensillum, made circular, and turned into a vibrating organ capable of detecting sound waves rather than mechanical stress. The ear may not have been developed primarily for listening to bats. There is some evidence from research on locusts that the ear may be monitoring wing-beat vibration frequency, giving the insect an in-flight checking system on the efficiency at which its wings are working.

As far as moth ears are concerned the most commonly studied group is the noctuid moths. The ear of a noctuid has only two nerve cells (compared with 17,000 in the human ear), not a great amount of neurological tissue to make sophisticated decisions on an attacking bat's location. The moth simply listens to see how loud the bat sounds. A faint bat is a distant bat, and a loud bat is too close for comfort. The moth can also discriminate from which side the bat is attacking. With two tympanic organs, placed one on each side of the body so that the body acts as a sound shadow, the moth can compare the intensity of sounds reaching each ear. If the bat comes from the right, the right ear will perceive the bat's call as louder than will the left ear and the moth can make the simple decision to veer off to the left. It has been demonstrated that moths with bat detecting ears have a 40% lower chance of being caught by a bat than moths without ears.

Together with Brock Fenton, James Fullard of Toronto University has continued the work on bat and moth interactions. Fullard considered it 'a lovely, almost a sensory ecology problem, where two animals that are using one particular source of sensory modality – acoustics and hearing – are using these in a rather sophisticated behavioural fashion'. Fullard observed that the first thing a moth will do, having detected a fast-approaching bat, is to fly away from the bat. If the bat is flying faster than the moth, and gets closer, the moth begins a series of aerial acrobatics. It will fly in large loops and circles to try to dodge the bat's attack. If you stand under a street-light at night, you can sometimes witness the manoeuvring. Most of the time the moths fly into the light but every now and again they will make a loop and dive rapidly towards the ground. As if from nowhere, a bat will swoop in. Sometimes it gets its prey; sometimes the moth out-manoeuvres the bat. If a bat gets to within a metre of a tiger moth, however, the moth emits a sound which seems to confuse the bat and it aborts its attack. If the bat keeps coming, the moth's last strategy is to fold its wings and drop. Often in the morning, where these aerobatic fights have taken place over water, the debris of the night is left floating on the surface. Moths that have dived to avoid the bats have hit the water, only to drown. Some bats specialise in picking off moths driven into the water by bats higher up (see Plate 11).

The most interesting of the moth's defences is the process of jamming. The moth is able to emit sounds that are thought to jam or interfere with the bat's radar. David Pye was able to record these jamming sounds in the laboratory at Queen Mary College. His early work on moth jamming in the early 1960s was with a neo-tropical tiger moth *Melese laodamia* from the West Indies. Using a tethered moth in front of a fan (in the fan stream the moth will fly continuously and silently), he challenged it with a train of pulses that simulatd the echolocation calls of an attacking bat. In response the moth produced a jamming noise which, when slowed down by about 32 times, sounded very much like a finger being rubbed back and forth along the teeth of a comb.

The moth makes these sounds with a modified scleroid on the side of the thorax. The plate has a series of little ridges and when it is distorted the ridges pop, much like a 'party-popper'. The sounds are very high frequency, in the ultrasonic range, above 20,000 Hz.

At first, it was thought that tiger moths might have their own echolocation system, using the buckling plate as the sound producer. Then it was considered that tiger moths, which are unpleasant-tasting insects, might be using the sounds as a warning signal to predators, much as those daytime animals that are protected by stings or distasteful venoms advertise their presence with warning colourations – bright black and yellow bands. And just as some hoverflies mimic the warning colouration of wasps, so too might perfectly edible moths mimic the sound of bad-tasting moths.

More recently, however, the work of Brock Fenton and James Fullard, in Canada, and James Simmons, in the USA, took a closer look at moth sound production, and put the sounds through a slightly more sophisticated method of analysis. It turned out that there was a remarkable similarity between the moth sounds and the calls being put out by the bats. The bat calls of interest are those produced in short staccato bursts or buzzes just prior to homing in on a target. They sound just like the moth sounds. This suggests that the moth is actually jamming the bat. The bat flies along, gets an echo from a moth, just like all the silent moths it has chased all evening, and just as it homes in, when it is maybe half a metre away, flying and diving at maximum speed, the moth, all of a sudden, throws a burst of sound at it. The bat, not expecting this strong signal, is confused, even startled. The bat, after all, must make sure it doesn't run into walls or trees, and on hearing this odd signal might think the best thing to do is to turn away. In experiments in the laboratory they do just that. Bats were trained to take a mealworm thrown into the air. When they became proficient catchers, tiger moth sounds were played just as the bat was about to catch the mealworm and the bat would veer off, leaving the mealworm alone.

So, how might the jamming work? The bat's brain is tuned to respond to a particular set of frequencies, i.e. those of its own echolocation calls and echoes. A series of filters ensures that the bat only hears and responds to those frequencies, and disregards irrelevant signals. In laboratory experiments researchers have tried to jam the bat's radar. They have trained bats to fly obstacle courses and have then played various sounds at them in order to confuse and disorientate them. White noise, with its broad band of frequencies, would naturally upset a bat at first, but the bat gradually worked out how to lessen the effect. It would reduce its flight speed, try to keep its back to the loudspeaker, and increase the intensity of its own echolocation calls. The experiments were gradually refined until the frequencies, intensity and temporal pattern used were the same as a bat would expect to hear in a returning echo as it approached a target. These sounds did confuse the bat.

The moth's sounds do not mimic the bat's call, but consist of the right frequencies and intensity to pass through the bat's auditory filters. This causes the bat to hesitate as the jamming sound interferes with its ability to process information in its echolocation system: a subtle kind of jamming. An analogy used by Brock Fenton is that of someone at a party, deep in conversation, who hears somebody on the other side of the room talking about him or her. The person's ability to concentrate on the conversation would be temporarily interrupted as his attention was distracted to the other conversation. The pulsed sounds of the tiger moth, similarly, interfere with the bat's ability to concentrate on the echoes and this causes it to break off the attack.

Moths, though, do not have the last word, for in the course of evolution, bats, it seems, have not given up that easily. Bats have counter-counter-measures to catch moths. If a bat raises the frequency of its echolocation call above, say, 60,000 Hz, or lowers it to below 20,000 Hz, then it lies outside the moth's optimum hearing range. A moth which is sensitive to a 25,000–50,000 Hz frequency range can detect a bat calling at this frequency at 40 metres. The same moth, when presented with a 100,000 Hz call, will only detect it at two metres; and similarly a 15,000 Hz call would be detected only at two metres. And some bats do just what would be expected – they emit their call at frequencies the moth is unable to detect. In Africa, Brock Fenton and James Fullard have found moth-eating bats that are putting out echolocation cries at very high frequencies. Such extreme forms of echolocation seem to be confined to the tropical bats where there are many insects from which to choose. Bats from Zimbabwe and Papua New Guinea have been found with very high frequency calls. In south central British Columbia in Canada, Fenton and Fullard have found a bat with an echolocation cry so low that moths, apparently, cannot hear it.

Clearly a moth, with its unsophisticated ear, cannot pick up every possible sound. It is best tuned to particular frequencies. All the bat need do is play around, in an evolutionary way, with its echolocation signal and the chances are that it will come up with something the moth cannot hear. But moths are part of the evolutionary tug-of-war, and they have adapted too. Although the structure and nervous system of moth ears is much the same the world over, James Fullard has noticed that sensitivity varies considerably. Moths in the tropics have to cope with a huge diversity of insectivorous bats and have ears sensitive to a much broader range than their more temperate cousins.

Moths are not the only flying insects to have ears. Another night flyer that may fall victim to bats is the green lacewing *Crysopa spp*. This delicate, green insect has ears located on a vein in its wing. Lee Miller in Denmark has discovered that the lacewing's ears are more sophisticated than those of moths and may be used also for communication between individuals as well as for bat detection.

Katydids, the bush crickets of North America, make their sounds by rubbing the wings together. Their chirpings are in the ultrasonic frequencies used by bats. Their ears, therefore, can detect bat sounds and there is some evidence to suggest that they are used in a secondary function to do just that. Researchers from the University of Florida are now in Panama looking at the interactions between katydids and bats. The katydids need to sing to attract mates, but their songs also attract bats.

The Fishing Bats

The fish-catching bats are 'FM' bats, using frequency modulated sweeps, much like the vespertilionid bats. If the fishing is poor, they are also known to take insects. More usually, though, they are seen to sweep low over water and take fish near to the surface. They don't dive like terns but have large clawed toes on the hind feet which they use to grab the fish. All the time they produce echolocation clicks. Whether the sound penetrates into the water is not known. It is more likely that they are detecting small disturbances as fish break the surface.

Don Griffin and Roderick Suthers brought one of the commoner fish-catching bats from Trinidad into captivity. It is known as the bulldog bat *Noctilio leporinus* from the overlapping lips which it uses to store chunks of fish. It is a large bat with very large hind feet, and flattened clawed toes to reduce water resistance. The bat flies along and, when a few feet from the target, dips the hind feet into the water and hooks a fish the size of a minnow. In an outdoor flight-cage, Griffin and Suthers recorded the sounds produced by the hunting bat. Throughout its flight the bat produced ultrasonic calls with plenty of 'FM' sweeps, indicating that it was concentrating on particular events rather than simply finding its way. The bats could distinguish food from other disturbances such as floating leaves or twigs or bubbles. The experimenters then put a piece of fish, about a half-an-inch across, on a wire which could be raised and lowered in the water. If the fish was out of the water completely the bats would scoop it up every time with ease. If the fish was placed just below the surface, so there were no ripples, then the bats were no better than chance at finding the piece. If the fish was kept below the surface but a piece of the wire protruded above by a couple of millimetres, then the bat quickly learned that food was to be found below the wire.

In the wild, there are often many of these bulldog bats fishing at the same time, on the same patch of water, and there is often a traffic problem. Griffin discovered that, when two fishing bats were on a collision course, both bats would use 'FM' sweeps and tag on an extra signal. This Griffin called 'honking'. The bats would veer away and avoid colliding. In this case, the echolocation call has been slightly modified to serve a communication function.

Social Calls

In addition to communicating with itself, an echolocating bat might also be communicating with others. Bats may be recognised by their echolocation calls. Just as an ornithologist can distinguish a robin from a wren by listening to its call, so too can a biologist who studies bats tell one bat from another. Why, though, should bats have different calls? Would it not make evolutionary sense for bats to converge on one call which would give them most information about a target? There is no obvious physical or acoustical reason that one call is better than another for detecting one kind of insect or whatever. An ornithologist looking at different call patterns in birds would suggest that the differences were for species recognition. Could it be that bats tell each other apart by their echolocation calls? A calling bat is giving away much information about itself; where it is located; where it is heading; whether it is chasing insects or not; what species it is; so it may even be identifying itself as a particular individual.

Robert Barclay, working with Brock Fenton in Canada, designed a series of experiments in an attempt to discover whether the calls that one bat uses to ask questions about its environment provide information for other bats. Does one bat listen to the echolocation calls of another, and if so, how does it react? Barclay carried out playback experiments with little brown bats. In an area where little brown bats were known to be feeding, an observer would lie on his back, underneath the playback loudspeaker, and count the number of bats that flew through the airspace above him and the direction in which they were flying. He was interested in whether they flew towards or away from the loudspeaker when different sounds, including those of little brown bats, were played. The observer had no idea what was being played, or even if a signal was being produced. When their own calls were played little brown bats would come towards the loudspeaker apparently looking for a concentration of food. Artificial echolocation calls, with the right frequency components and the right durations, produced the same reaction. Little brown bats have a frequency modulated sweep from 80,000 Hz to 40,000 Hz in four thousandths of a second (four milliseconds). If the same sweep was played but with a time-base of 50 thousandths of a second, the bats paid no attention. The correct call played backwards, so that the sweep went up in frequency, was again ignored. Little brown bats, then, Barclay and Fenton concluded, use the calls of others of their own species to locate food. It must be remembered that a little brown bat detects a target, by echolocation, at a range of only two metres. If it listens out for the calls of others it could extend its effective range to, say, 50 metres. There is clearly an advantage for one bat to listen to another.

Martha Leonard, also working with Brock Fenton, studied another species of North American bat, the spotted bat *Euderma maculatum*. It is a large bat, 55 millimetres long, with large pink ears, jet black fur and big white spots on the back. Spotted bats do not feed in groups, but tend to be solitary. Leonard carried out a similar series of playback experiments with spotted bat calls and discovered that the bats responded in one of two ways. Some would fly to the speaker and attack it. Others flew away. Leonard and Fenton concluded that the echolocation calls of spotted bats not only ask questions about the environment, but also serve to space out individuals.

Little brown bats feed in the evening over water where insects, such as mayflies (Ephemeroptera) and caddis flies (Trichoptera) often occur in patches. They could not defend such a concentration of insects, nor indeed do they need to, for there is such a large amount of food in a patch that poaching by other individuals would have little effect. The spotted bat, on the other hand, hunting in ponderosa pine forest, has a food source which is well dispersed and it would be important to keep the density of bats down.

But information passing between bats may not be confined to food gathering. Echolocation calls may also help in determining the place where a bat comes to rest. In Zimbabwe, Brock Fenton looked at the roosting habits of a species of bat known locally as the house bat, although the particular group he was studying did not roost in houses. After following the group for some time, Fenton noticed that they would find a fresh roost each night. He suggested that they were homing in on the echolocation calls of others to determine in which tree they would all roost. Echolocation calls would be an easy location cue to gather the group together and, for bats, knowing where all the other bats intend to roost is an advantage. Bats suffer from an energy conservation problem. They are small animals, have a large surface area, and lose heat rapidly. To keep warm, they cluster together. Their collective body heat raises the temperature of the roost, and finding a communal roost is important to the individual bat.

Bats also have other types of calls. Bats often congregate in huge nursery colonies. As many as a million female bats and their offspring, usually one a year or for some species twins, may be found packed into a cave in tropical latitudes. To the casual observer it looks like chaos. A million screaming babies hanging from the cave roof must be found and looked after by a million mothers. It was suggested at one time that bats' mothers do not attempt to find their own baby but simply give a drink to any infant that wants one. That view of bat infant care has now changed. Chris Thompson, another student working with Brock Fenton, carried out some work on a nursery colony of little brown bats in eastern Ontario. She found, contrary to what had previously been thought, that mother bats were very good at finding their own offspring. When the mother arrived in the cave, her baby would be triggered to cry in response to her echolocation call. The baby produced specialised vocalisations, called isolation calls, which the mother used to locate her baby. The mother herself then emitted a quite distinctive double note call which again set the baby off. In addition to sound, smell and vision are also important in mother/baby interactions. The mother first uses spatial memory. She remembers where she left her baby in the cave. The isolation call gives her the general area, in which there may be still a hundred or so babies. Smell would then be used to locate the individual infant. The babies were not as discriminating. They would try to sneak a drink from any mother that was passing, and get bitten for their trouble.

Little brown bats have a copulation call. The male produces a 'weep-weep' sound with a rising frequency sweep. During mating, female little brown bats struggle and it is thought that the male's call is given to pacify them.

The social behaviour of the 'whispering' leaf-nosed bat *Carollia perspicillata*, a fruit-eating bat from Central and South America, has been under the watchful eye of Fran Porter, of Washington University, St Louis. She has found that these bats rely heavily on acoustic signals for social communication. Flying adults, and flightless infants throughout the first three weeks of life, constantly scan their environment with soft click-like 'FM' pulses.

During courtship, male bats emit 'whines' as they approach females and 'trills' during mating. 'Warbles' are given when animals are greeting, particularly between harem males and their females. In times of threat or distress the bats give raucous 'screeches', for example in aggressive chasing bouts between males, and in mid-air collisions. Males appear to 'screech' at harem females in order to round them up and bring them in closer together in the roost. Males would also call to mothers, separated from their babies, to encourage them to find and retrieve their offspring.

When Fran Porter removed infants from the nursery, an interesting series of events took place. Porter played back recordings of the removed infant and watched the mother's reaction. During 16 experiments only six mothers flew to the playback loudspeaker. They seemed uninterested. On the other hand, the harem male, with whom the mother roosted, reacted in one of two ways. Either he would fly directly to the speaker playing the infant's calls, or he would fly to the correct mother and poke and screech at her. Even neighbouring males would fly to the mother and start to wing-poke her, and occasionally females in the same roost would approach and flick their wings in an agitated manner. Both the neighbouring male and the harem male would continue screeching at and poking the mother from either side until she finally flew towards the speaker. Females would only attempt to retrieve their own offspring.

Fruit Bats

The 150 known species of fruit bats or flying foxes (Megachiroptera) have a rich repertoire of sounds, including courtship calls and mother/baby calls, but with one exception do not echolocate. Fruit bats have superb night vision and do not chase about after insect prey; they eat fruit.

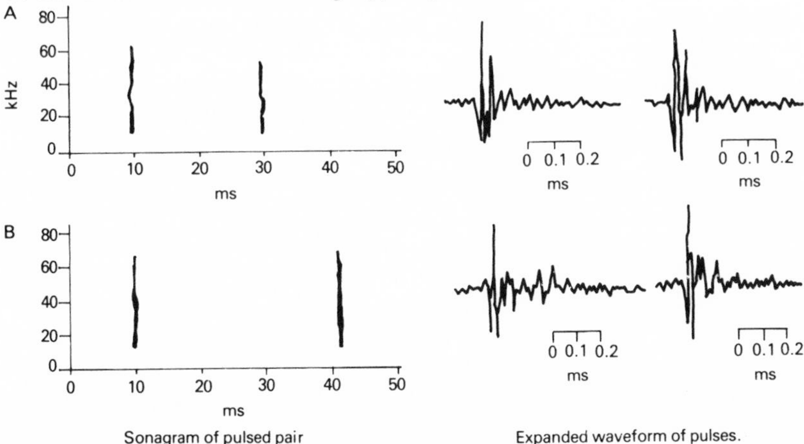

Fig. 38 Echolocations clicks of: A. *Rousettus aegyptiacus* and B. *R. amplexicaudatus*.

The odd bat out is the *Rousettus* fruit bat, or dog-faced bat, a genus with four known species, which is found from Africa through to the Far East. *Rousettus* fruit bats echolocate, but in a quite different way from the small insect-catching bats. Unlike the small bats, which produce their sounds vocally with a voice-box, *Rousettus* clicks the tongue. Each click (known in German as a *Zungenschlag*) which sounds to us as a single click, when analysed, consists of two separate clicks about one-fifth of a second apart. As far as is known, these fruit bats do not use the sounds to orientate in the dark at night; they have night vision for that. It is thought that they use the tongue clicking for roosting in the safety of totally dark caves during the day. Echolocation with clicks is a little like a blind person tapping a stick to listen for echoes. Oil birds and cave swiftlets have exploited echolocation by clicking during cave roosting in a similar way.

11
ANIMAL INSTRUMENTALISTS

The Crustaceans

Crabs, lobsters and shrimps are not the silent denizens of the sea that they might at first appear to be. Ten or 20 metres below the surface of a tropical sea there can be bedlam. Snapping shrimps, in huge clattering choruses, and quintets of rasping crabs and lobsters make the ocean a very noisy place indeed.

Lobsters

One crustacean that has been studied in some detail by John Mercer, director of the Shellfish Research Laboratory of University College, Galway, is the gregarious spiny lobster *Palinurus elephas*, to be found on the sea bed off the rocky Atlantic coasts of Europe. Sound communication, it seems, is an important part of a spiny lobster's life.

Spiny lobsters make a rasping sound, rather like a finger being rubbed along the teeth of a comb, produced when a serrated pad at the base of the antennae rubs on a projection on the rostrum, the pointed front end of the carapace or shell, between the eyes. The rostrum itself acts as a sound box to amplify the sound, which is audible up to 50 metres away. Lobsters do not have ears to receive the sound. On the antennae, and on other parts of the body too, there are minute tufts of hair. Some hairs are chemosensory while others respond to water displacement.

The sounds made by lobsters serve a variety of functions. There is a general conversational call that keeps the whole group together, and indicates that it is safe to go about foraging and grazing. In a population of spiny lobsters, which can be anything between 100 and 1,000 individuals, there is a constant background chitchat of rasping sounds being produced at a fairly low level. If danger threatens, for example, a shark or conger eel looking for a meal, the conversational call stops instantly and the warning call is given. This is the same rasping sound but given at a higher frequency rate. The higher frequency call encourages the entire population to rush for shelter in cracks in rocks and under overhangs.

There is a close relative of the European spiny lobster living off the coast of South Africa. The lobster often shares the same hole as a moray eel. The moray eel does not prey on the spiny lobster but the local octopus does. The moray eel preys on the octopus. When an octopus attacks a lobster, the lobster gives its rapid warning call. The moray eel, on hearing the sound, shoots out of its hole and catches the octopus.

One situation where animals do not want to escape is in courtship. Prior to mating, sound plays a key role for spiny lobsters. But it is not the male that makes all the noise – a prima donna takes the underwater stage, for lobster mating calls are performed by the female. Her mating call can attract males in a circle around her for about 50 metres. The female lobster sits on a rock and stridulates. All mature males head towards her. As the circle tightens the males begin to fight for the privilege of mating. One male reaches the female and she stops calling. The rest, as though a switch had been flicked, immediately lose interest and go about their daily tasks. The victor and the female then engage in a more elaborate courtship procedure. They caress antennae. The female releases pheromones into the water, and once this has happened the male is programmed to mate with her, no matter what.

Female choice in lobsters is unlike that in other animals. The male chosen is not the biggest and the best, although size is still important; only lobsters of equal size mate. On the carapace of the protagonists there is complementary sculpturing, a lock-and-key mechanism that only allows animals of a similar size to pair. This prevents lobsters of different species from mating.

Crabs

Crustacean conversations are not confined to lobsters. Ghost crabs *Ocypode spp.* defend their muddy burrows with territorial calls. These are produced with the aid of ridges on the inner surface of the crab's large claw. Calls, amplified by the horn-shape of the burrow, also serve to attract females to the males' courtship sand pyramids.

Varieties of the Chinese land crab, which produces sounds by rubbing the claw or pincer against a series of ridges on the body, even have local dialects. The stridulatory ridges of crabs from geographically different areas have varying degrees of coarseness.

In 1894, Anderson wrote about tropical, swift-moving, air-breathing land crabs *Ocypode ceratophthalma* which rasp a ridge on the larger pincer against a ridge on the second joint of the same limb:

> 'One bright hot sunshining morning in November, as I walked along the shore of Bingaroo, one of the Lakadive Islands, which is occasionally visited by the inhabitants of other islands of the same atoll, I was

surprised to hear a loud creaking noise, that appeared to proceed from the edge of the scrub jungle that covers the island. At first I imagined it must be caused by frogs, so perfectly did it resemble the croaking of these animals. However, on tracing the sound to its source, I discovered that it proceeded from the burrows of the Ocypode crabs which here fringed the beach at high-water mark. These burrows are frequently, in coral sand, very wide at their mouths (six to eight inches), and then taper gradually downwards, so that they act as excellent resonators. The cause of the stridulation of the crabs was by no means apparent, the animals were all lying hidden in their burrows, and several were croaking at the same time, as if in concert'.

Today, it is generally thought that the stridulations advertise occupancy.

Shrimps

The noisiest of marine crustaceans is, perhaps, the snapping or pistol shrimp *Alphaeus spp*. By dislocating an enormous claw, half as big as its own body, which it extends in front, the pistol shrimp can produce a very loud report. Like dolphins, these shrimps may be using high intensity sounds to stun prey. They will stalk small fish and when close to them will fire the claw. The fish tips over and the shrimp catches it. The sound is so loud that if shrimps are held in an aquarium in which the glass is scratched, and they pull the trigger, the glass is likely to shatter!

Mantis shrimps *Gonodactylus festai*, studied by Ray Caldwell of the University of California at Berkeley, defend their cavities with visual, tactile and acoustic displays. The mantis shrimp does not have a specialised sound producing apparatus like the snapping shrimp, but strikes itself against its cavity wall, producing a loud click which can be heard two metres away. If an intruder approaches the cavity entrance, the resident will strike its cavity many times. Caldwell believes the sound discourages intruders from entering the cavity and proclaims that the cavity is occupied.

The Spiders

After emerging from the sea, and reaching the land, the hard-bodied invertebrates continued to exploit sound as a means of communication. One group of animals not usually associated with love songs and territorial calls is the spiders, but spiders, it seems, make a variety of sounds.

George Uetz, of the University of Cincinnati, and Jerry Rovner, of the University of Ohio, are keen on spiders, particularly the noisy ones. They are behavioural ecologists interested in the adaptiveness of behaviour, in particular how the courtship behaviour in two species of wolf spiders functions to keep the two species reproductively isolated. Using sophisticated ultra-sensitive recording equipment, they have been able to pick up the faint sounds passing from one spider to another. The sounds are, in fact, inaudible to the human ear, but Uetz and Rovner have been able to identify 27 families of spiders that are capable of making and using sound for communication. They have managed, so far, to record about a dozen.

There are early reports in the literature of people hearing sounds from some of the louder spiders, for instance from certain wolf spiders which can be heard if the area is extremely quiet. They drum leaves making a quiet purring noise. After these first observations in the wild, Uetz and Rovner have brought spiders into the laboratory and investigated the sounds being made.

To tape the sounds of spiders, Uetz and Rovner had to devise their own specialised recording techniques. They use a vibration pick-up, called an accelerometer, which is mounted on a piece of stout paper placed on a very heavy laboratory table, isolated from vibrations in the rest of the room. They have tried a variety of benches. The favourite is a laboratory balance table-top made out of heavy marble with the legs mounted in buckets of sand, and surrounded by foam-rubber blocks. The sound vibrations in the stout paper are picked up by the transducer. The signal is pre-amplified and recorded, but background noise with such low level recording is a problem. The tapes are cleaned up by DBX noise reduction so that the spider sounds stand out from the background noise.

So, in a quiet room with a sheet of stout paper on which to place the subjects, a sensitive vibration detector to pick up the sounds, and a sophisticated tape recorder on which to record them, the Cincinnati researchers have been able to analyse spider sounds and substrate vibrations and relate them to particular patterns of behaviour. Spiders, like many other animals, were found to use sounds during courtship. A spider is induced to court by allowing the female to run back and forth across the paper laying down silk lines. There is usually enough pheromone in the silk to stimulate courtship. The male is then confined to the paper and his sounds recorded.

Male spiders produce a courtship buzz, both to identify the species and to impress the female. Without hearing the sound the female is unlikely to be sexually receptive, which could prove fatal for the male; the female might eat him! The female spider picks up the sound with sensitive hairs in receptor organs in the exoskeleton of the leg. Uetz was able to separate spiders so that the female could either hear the sound and not see the spider, or see the spider and not hear the sound. He found that the female would not respond without the sound signal.

Like other arthropods, spiders do not make vocal sounds; they are instrumentalists. The wolf spiders, studied by Uetz, have a stridulatory file-and-tooth system on the joints of the pedipalp, the two leg-like organs below the spider's mouth. In the male the pedipalps are also used for sperm transfer. In the case of wolf spiders, the vibration of the file-and-tooth system is conducted to the substrate, a leaf for example, by stout spines at the tip of the pedipalp. The vibration is then transferred across the leaf surface to the receiver.

George Uetz has been studying two species of wolf spider that are, in their outward appearance, identical. They are both about one centimetre long and coloured black and brown with a pale grey or white band down the back. In one species the male has the front legs decorated with dark brown/black coloured tufts – the only visible difference. It is only when hearing the sounds that they can be told apart – but you have to listen very carefully; they produce sounds just on the threshold of human hearing (although some researchers believe the airborne sounds to be spurious, the main signal being transferred via vibrations in the substrate).

The courtship of one species, *Schizocosa ocreata*, involves the male tapping the front pair of legs on the ground, at the same time stridulating with its pedipalps and following a trail of pheromones laid by the female. It makes a kind of purring sound. Occasionally, the spider stops and arches, waving one or both front legs in a visual display to the female.

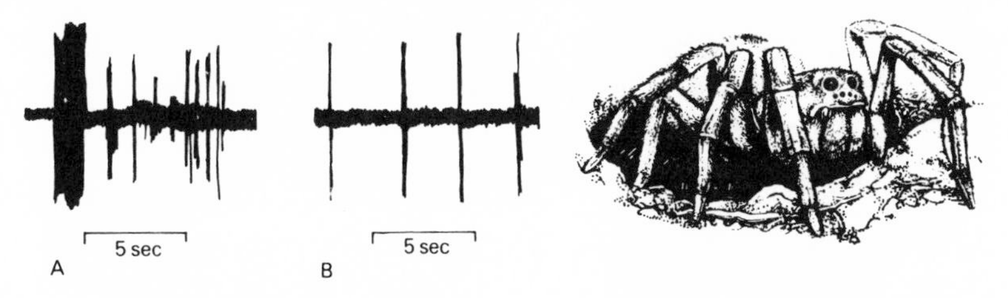

Fig. 39 Oscillograms of male wolf spiders. A. *Schizocosa ocreata*. B. *S. rovneri*.

The other species, *Schizocosa rovneri*, also follows the female's trail, but carries out a different piece of behaviour. Every time it stops it begins to 'bounce', lowering its body between its legs every three seconds. Each 'bounce' is accompanied by stridulation of the palps. The sound can be heard as a quiet buzz, presented in short bursts.

The frequency range of wolf spiders is narrow and low, at about 250 Hz. It is the temporal patterning of the sound which is important. In the two wolf spiders studied by Uetz, for example, one produces sounds in very short bursts, while the other has long chains of stridulations. The female spider appears to make no sounds. She does not have the stridulatory organs of the male. Her response to the male's advances involves a body movement dance. Uetz has a graduate student looking at the possible discriminating abilities of female wolf spiders. Individual male spiders, perhaps, have slightly different sounds from those of their rivals, so that one male might have a better chance of mating than another male of the same species.

Many other spiders make sounds, but with stridulatory organs on different parts of the body. It is likely that these animals evolved sound producing organs by moving body parts together, for example, the rubbing of the fangs or the brushing of a leg against the abdomen. Some wolf spiders make a percussive noise. They simply tap the pedipalps against the substrate. George Uetz suggests that the pedipalp joint stridulation probably evolved from a tapping behaviour. One European spider produces vibrations across water surfaces. It taps a hard, sclerotised plate at the bottom of the abdomen, on the substrate. Whether this is achieved by tapping a floating leaf or by directly tapping the water, is not clear.

Some of the more familiar garden spiders also create ripples, not in water or on leaf surfaces, but across the web. The spiders have stridulatory organs which are used to produce a web-borne vibration. They produce sound by flexing the legs against the abdomen. Some species actually pluck the web.

Most spiders live alone. They are cannibalistic. There is, however, a social spider, *Mallos gregalis*, that lives in colonies. The colony catches food collectively. The colonial web is a sheet which is so designed that vibrations from buzzing prey insects landing on the web differ sufficiently from vibrations between colony members so that the two cues are not confused and a social lifestyle is possible.

Jerry Rovner and Friedrich Barth, of the Goethe-Universität, Frankfurt, have discovered some large, tropical sparassid and ctenid spiders that produce

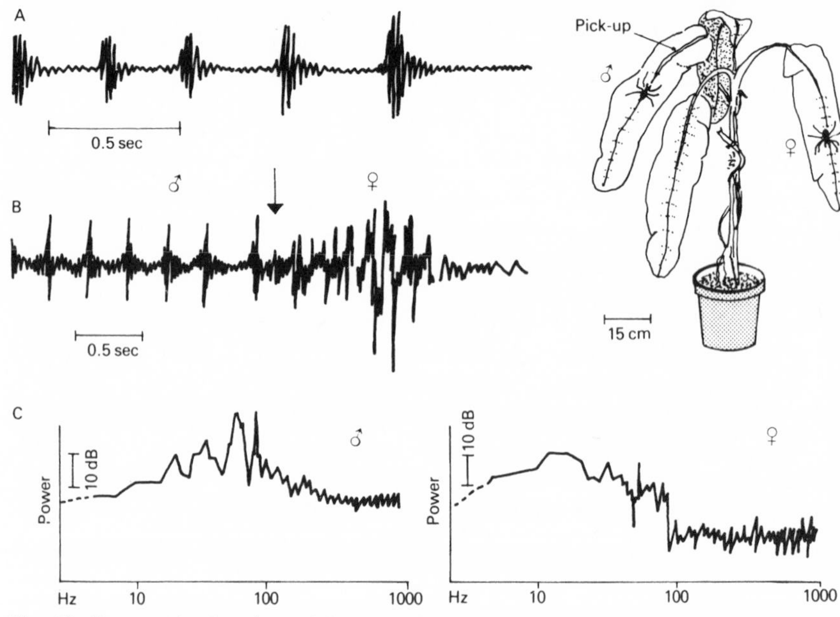

Fig. 40 Communication through banana plant by Cupiennius. Male begins courtship on pheremone-covered leaf-blade, and selects correct petiole to reach female on other leaf. A. Male pulse train. B. Male pulse followed by female on another leaf as received by the pickup. C. Power spectra of male and female signals as recorded at the pickup.

sounds by whole-body vibration. There are no specialised organs. The spider vibrates its entire body very rapidly, and shivers, producing a humming noise. One ctenid spider, the nocturnal wandering spider *Cupeiennius salei*, transmits the vibration through the leaves of banana plants. The male and female signal back and forth, both producing the vibrations. The entire courtship procedure is initiated by pheromones deposited on the leaf by the female.

Spiders also have defensive sounds. Some species of tarantulas have stridulatory organs that produce a kind of hissing noise that resembles the hiss of a snake. And there are calls that males direct at other, possibly rival, males. Jerry Rovner has been working with some large species of wolf spiders in the genus *Lycosa*. He has found that male wolf spiders produce sounds in aggressive interactions with other males. The sound-producing organ used is the same as that used for courtship calls, indicating that spiders can use the same structure to produce a variety of calls. In wolf spiders there is a pattern for impressing females and another for frightening off males.

The Insects

Another group of invertebrates noted for their sound production is the insects. Just as there is incredible diversity in insect size and shape, so too there is a wide range of sounds and sound production mechanisms.

Cicadas

On a hot sultry summer's day in southern Europe, the air is likely to be filled with a chorus of cicadas (Cicadidae). Male cicadas sit high in the branches of trees, calling loudly in an effort to attract a mate. The song can be heard, sometimes, half a mile away. Once a single cicada starts to sing, the entire population starts up in an amazing clicking chorus, each one in synchrony with its neighbour.

Cicadas produce their sounds with a pair of tymbals. These are hard, sclerotised plates on either side of the thorax, which are buckled in and out very rapidly by the contraction of powerful muscles. When the muscle contracts the plate buckles in and makes a click; when relaxed the plate buckles out and clicks again. By raising and lowering a covering over the tymbal, the male cicada can make his song louder or quieter. One species of cicada from the Seychelles, studied by David Aidley of the University of East Anglia, is known to make as many as 1,000 clicks per second. Others are less ambitious and produce anything between 200 and 600 clicks per second.

The basic cicada system works by contracting the muscle once, thus producing two clicks – an 'in' click and an 'out' click. There are limits as to how fast the muscle can contract and relax. If a muscle contracts one hundred times a second (a very fast rate for muscle contraction) then with the in-click/

out-click system the bug can produce 200 clicks per second. This click rate can be speeded up if the muscles on opposite sides of the body contract alternately, producing a rate of 400 clicks per second. But an even higher frequency of clicks is desirable for some cicadas. This is achieved by having a series of ridges on the tymbal, so that the tymbal buckles in a series of steps, each making a click. A cicada could then produce four or five clicks for every inward buckling of the plate, and a further four or five with the outward buckling. A cicada with a modest 80 contractions per second would, by alternating the sides, achieve 160 inward and outward movements of the plate per second. With four ridges on each plate the 80 contractions per second cicada can turn itself into a 1,200 clicks per second insect.

There is an alternative method for getting high click rates. It is similar to the physiological mechanism used by flying insects, particularly mosquitoes, for activating their wings. The high-pitched whine of a mosquito is produced as its wings beat at a rate of 500 beats per second. No normal muscle can contract and relax in this short time. The way a muscle works is that it receives a nerve impulse, the muscle is switched on, it contracts, switches off, and then relaxes. For the next contraction a fresh nerve impulse is needed, and so on. In the special flight muscles of mosquitoes the muscle is switched on all the time by a succession of nerve impulses. During this period the muscle itself oscillates, contracting and relaxing, and can do so at a much higher frequency than if it was receiving separate nerve impulses. A cicada from the highlands of Kenya, also recorded and studied by David Aidley, is now known to use this oscillating muscle system in order to get very high rates of clicking.

On the receiving end, the ear of the cicada is interesting. It has a tympanic membrane or eardrum, and attached to the membrane, at one point, are about 2,000 receptor cells. This is a puzzle. Firstly, why should an animal like a cicada need 2,000 receptor cells in its ear? The males make such a noise that a female would not need so many receptors to detect them. Secondly, why are they all attached to one part of the membrane? If the cells were attached to different parts of the membrane it might make sense – the membrane might move differently with different frequencies. At the moment, this remains a mystery.

The cicada's sophisticated sound producing and receiving apparatus is designed to do two jobs – first the calls bring together the male cicadas into enormous singing groups, and second, the singing males attempt to attract the females into their midst. Ken Hill, studying an Australian cicada, has discovered that the female flies towards the centre of a group of bawdy, singing males, and, like a moth to a light, circles the area. At some stage she descends into the tree to join the party, after which she doesn't move. All around her males are singing. Then she begins to produce a 'ticking' sound by flicking her wings. Maybe it is the airborne sound, or maybe she is releasing a pheromone, but the males gradually move towards her. Each time she makes a 'tick' the males advance. When she stops, they stop. Eventually one male reaches her and the final stage of courtship ensues. How, or even if, she chooses a mate is not known.

For each species of cicada there is a different click rate. In order that the 1,500 known species of cicada mate with the right partner it is essential that there be some form of species recognition. Songs fulfil that function. Cicadas vary their clicking in a variety of ways – some have continuous sets of clicks; some have clicks all at the same amplitude; some vary the amplitude; yet others change the pattern of clicks.

Clearly, the more clicks an individual is able to make the easier it is to vary the signal. One of the reasons why cicadas have been evolving higher and higher click rates, suggests David Aidley, is to make their individual songs quite distinct from those of other species.

Crickets and Grasshoppers

Brian Lewis, an insect physiologist at the City of London Polytechnic, studies crickets (Gryllidae), bush crickets (Tettigonidae), and grasshoppers (Acrididae). He is interested in how animals recognise their own species, how a sound signal is tailored to provide that information, what goes on inside an animal to decode that signal and, having heard the signal, how the animal locates where it is coming from. Insects, he feels, provide a good but relatively simple model with which to investigate some of these questions. They are good for species recognition studies because the songs can be chopped up and analysed. They are not so good for sound location studies because most insects are so small and use, at least crickets tend to use, sounds of rather long wavelengths compared with body size.

The problem of species recognition involves the way sound is produced, its structure, and the way the receiver's ear and nervous system extract the information. Sound, in insects, has two obvious characteristics – temporal patterning and frequency content. Either or both of these parameters could carry species information. Frequency tends to vary as much within a species as it does between species, so Brian Lewis suggests that little species information is likely to be carried in that way. This does not mean that the frequencies used are not important. Each insect ear is tuned to receive particular frequencies and so will function best at those frequencies. Temporal patterning, on the other hand, is quite distinct. Different species of insects occupying the same habitat will have quite different sound patterns.

In the crickets and grasshoppers, there are two general patterns of sound production, which can be varied enormously between species. The more complicated method is that used by grasshoppers. The hindlegs are rubbed against the wings. Here there are two sound sources, the two legs, which could be moving together in phase, out of phase, sometimes in and sometimes out, to produce complicated song structures.

The simpler method is that employed by crickets and bush crickets. Here the two wings are rubbed together. This gives a single sound source. Sound is produced when the wings close but not when they open. There is a period of sound followed by a period of silence. The duration of the sound will depend on the speed of closure of the wings, and the periods of silence on the speed of opening. Again there are a variety of combinations which make different patterns. There may be two periods of sound, for instance, followed by a long period of silence, or there may be four or five sound periods with a speed change in the middle so that there is a slow start and a fast finish to each sequence or chirp. The packaging of the chirp is species specific.

The sound is generated by a tooth-and-plectrum system. On the edge of one wing is a toughened region, the plectrum, which runs against teeth on the underside of the other wing. Every time the plectrum hits a tooth there is a sharp click. The sound energy is fed to an area of the wing, which in the crickets is called the 'harp' and in bush crickets the 'mirror frame', that actually transmits the sound, much like the diaphragm in a loudspeaker. The size of the 'harp' or 'mirror frame' determines, like the length of a violin string, the frequency of the emitted note. If the rate of plectrum-tooth strike is the same as the 'harp' or 'mirror frame' frequency then the sound is said to be a pure, resonant song. This is the case in crickets. If the tooth strike is different then the sound is broad-band and rougher, with lots of harmonics. This is so in most of the bush crickets. The greater part of the bush cricket song is ultrasonic. We can hear only the impact-sound of the plectrum on the tooth.

Crickets usually have three different songs. There is the 4,000–5,000 Hz calling or proclamation song used by the male. He may be establishing a territory, attracting a female or maybe both at the same time. This is often a simple sequence of a few syllables in every chirp, which is delivered for long periods of time. When the female approaches and is within visual range the male changes to a second song, the courtship song. The syllables are shortened and the wings are held in a different position so that the frequency of the notes goes up. The female is aroused by the change in delivery rate and frequency. Different receptors in the female ear could be excited. The third song type is the longer-chirp aggressive song, which a male gives if another male approaches too closely. It is accompanied by an aggressive visual display with the jaws open.

Why, though, should insects start to sing and, having started, why do they stop? The motivational state of insects is barely studied, but there are some observations that give us a few clues as to what is going on.

Crickets, in general, sing at specific periods during the day; some sing from dawn to noon, others from noon to dusk, yet others sing at night and some just before dawn. They will not sing, however, until they have a spermatophore, that is, until they are ready for copulation. Experiments were carried out to ascertain why. Was it a mechanical feature which produced the stimulation, or a chemical substance produced by the sperms in the spermatophore? A spermatophore was removed, ground up, and the solution reintroduced into the cloaca. The animal refused to sing. The spermatophore was removed once again and an intact one replaced in the cloacal region. The animal began to sing.

The insect will sing for long periods using vast amounts of energy. If a female does not approach the male, the spermatophore is reabsorbed and the insect stops singing. When the next spermatophore is produced it starts to sing again.

In crickets, sound is now almost exclusively used for courtship and the attraction of a mate. There is some degree of territoriality. Males will space themselves out and, although the geographic distance might not be the same, the acoustic distance is always the same. Individuals place themselves where a rival's song is heard at about 40–50 dB, if the intensity at the source is 100 dB. If there is a big bush in the way a male might be geographically nearer, as the bush shadows the sound. Females appear to recognise these territories and are able to identify something in the signal of the singers which enables them to choose one male rather than another. Females have been seen to walk through the territory of one singing male to reach another. As yet researchers have not been able to identify any difference between the songs of the two rivals.

After visual contact with the female, the calling song changes to the courtship song; the female comes closer, and the male makes antennal contact. After a short while, the male lays his antennae across the female's back. She stops and turns her back to the male. The male climbs on top, stops singing the courtship song, and copulation takes place. After copulation the male gets down, but waits around for a while with his antennae on the back of the female. If he moves away too rapidly the female will eat the spermatophore. He therefore calms the female until the spermatophore is absorbed. He then moves off, not singing again until the next spermatophore is ready.

Sexual behaviour in the bush crickets is less well studied. Living in dense undergrowth, and singing at night, these insects are difficult to observe, and do not like to be disturbed. The pattern seems similar to that of crickets, except that there is no courtship song, only a calling song. The female approaches a singing male, she stands still, the male climbs on her back, they copulate and he moves away.

How, though, does a singing male know if it is a female or a silent male approaching? The implications are that chemical behaviour is involved. The touching of the antenna appears to establish sexual identity. If a singing male approaches, the aggressive song is sung. Males also sing this song if a female attempts to leave too soon after copulation.

There has been some speculation about the way in which insect songs are acquired – are they innate or learned? Brian Lewis thinks the case for learning must be minimal. The majority of insects die off at the end of the breeding season. Eggs are laid, maybe, in the ground and the first offspring to arrive will never have heard adult song, yet they are able to produce the song as soon as they become adult.

There is some evidence that the pattern of song improves with practice. In the penultimate stage of the development of crickets (the final instar), although there are no wings, the wing muscles of the nymph move in synchrony with the song pattern. The movement is uneven at first, but improves during the development of the instar. By the time the animal is adult, it can produce accurately the full-blown song.

The genetic basis of sound production has been studied very little. Two closely-related cricket species have been crossed and the songs of the offspring examined. It turns out that they sing the song of the paternal species and the females prefer this song. So in crickets, at least, it looks as if the male genetic component for song production is dominant over the female contribution. However, in grasshoppers, where some of the females do sing, the offspring may produce the song of one parent, but prefer to move towards the song of the other parent. This indicates, perhaps, two mechanisms which, although related, are not tightly bound genetically. There is a genetic control for sound production, on the one hand, and a genetic control for recognition of sound on the other.

The nervous control of song production has still not been worked out. Neurones have been identified as active when sound is produced, firing in synchrony with the song pattern. It is not known what controls the pattern. Electrodes placed in specific areas of the 'brain' can stimulate an insect to sing. The 'brain', though, does not seem to be absolutely necessary because if the head is removed animals will often go on singing.

As far as hearing and species recognition is concerned, it looks as if temporal patterning is the main factor, although it is not that simple. The rate of syllable production is similar in most crickets. It is the duration of the sequence of syllables, the chirp length, that differs. Experiments, therefore, have been directed at searching for preferred frequencies. The response of the ear to pure tones has shown that the best frequency response of the ear is the same as the dominant frequency of the song. Apart from that, little is known about how the ear and the central nervous system recognise the species pattern.

Brian Lewis has been carrying out experiments with an Australian bush cricket *Teleogryllus oceanicus*. It has a song with a slow start and a very fast second phrase. To his surprise, he found that some of the central nerve fibres are inhibited by the main carrier frequency of the song (4,500 Hz). Lewis was puzzled. He looked more closely at the frequency content of the song.

Until then it had been reported that the cricket had a dominant frequency of 4,500 Hz. But no-one had looked above 16,000 Hz. Lewis did so, and found harmonics at 16,000 Hz, at 32,000 Hz, and another at 64,000 Hz, with various side bands in between. This cricket turned out to have a very complex song. Lewis tested the nerve fibres once more, this time using a series of simulated songs. He followed the temporal pattern of the natural song, but used different frequencies.

The response to 4,500 Hz alone, even with the correct song pattern, was poor. If 16,000 Hz was played the response was good, but the patterning was ragged. A mixture of 4,500 and 15,000 Hz, in a simulated song pattern, was coded perfectly. What Lewis and his colleagues thought was happening was that the 16,000 Hz frequency band was giving a very strong excitatory response, which was, in turn, inhibited by the 4,500 Hz frequency band. There is a balance of excitation and inhibition.

In the European field cricket *Gryllus campestris*, Lewis has not found the same type of central units. Its song has a dominant tone at 4,850 Hz but, unlike the Australian cricket, it responds quite happily to that frequency, and even to other frequencies as well.

One of the major questions in which Brian Lewis has been interested is, how do insects locate a sound source? For a cricket, standing on the ground surrounded by undergrowth, or a bush cricket in a bush surrounded by leaves and twigs, this could be a serious problem. The situation is different for each of these two types of orthopterans. Crickets sing at comparatively low frequencies, around 4,500 Hz (about ten centimetres of wavelength). The distance between their two ears on their front legs is likely to be about one centimetre, so there are unlikely to be any sound shadows between one side of the body and the other. Bush crickets use much higher, ultrasonic frequencies with wavelengths measured in millimetres. Again the distance between the ears on the front legs is about one centimetre, so there is likely to be some sound shadow. Consequently, these two groups of insects adopt different ways of locating sounds.

First, then, the bush crickets. In 1974, Brian Lewis showed that the main port of entry for airborne sound waves was not directly from the outside onto the tympanic membranes on the leg, but rather through a very large opening on the prothorax, an enlarged and modified respiratory spiracle. The trachea associated with it has become separated from the rest of the respiratory system, and is known as the acoustic trachea. It descends from the spiracle down into the leg to the back of the tympanic membrane. Looked at in longitudinal section it is an exponential horn. This acts as an amplifier. So the sound waves enter the spiracle, are amplified something like ten times in the acoustic trachea, and then hit the *back* of the 'ear'.

Lewis looked at the directionality of the system by placing a bush cricket in an anechoic chamber, and moving a sound source 360° around the subject. At appropriate frequencies he discovered a 20–30 dB difference between sounds received by the two ears. With the spiracle blocked, the difference disappeared. He also found that most of the left/right difference is due to sound shadowing by the insect's body, although there is some discrepancy caused by peculiarities of the tracheal system itself. The spiracle and acoustic trachea function in the same way as the mammalian pinnae, the large external ears.

The distance between the ears of crickets is about four or five times smaller than the wavelength of their calls. There is likely to be little sound shadowing

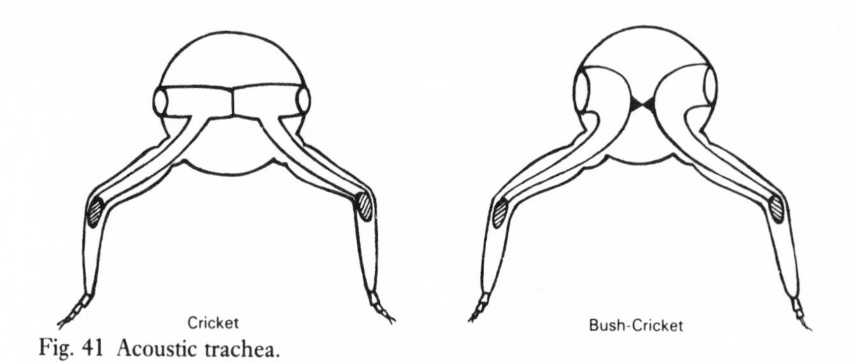

Fig. 41 Acoustic trachea.

(about 3 dB left/right difference). How then do crickets achieve sufficient difference between the ears to be able accurately to locate sounds? They do this, says Brian Lewis, very cleverly; they too use an acoustic tracheal system.

Like the bush crickets, the cricket acoustic trachea opens on the body surface. It differs in that the trachea from the two sides run towards the centre of the body and join, before branches run down into the leg and the back of the tympanic membranes. There is, then, a direct acoustic pathway between the membranes of each side – up one leg, across the body, and down the opposite leg. Sounds can travel in that internal cavity. At the same time, sound travels unimpeded around the outside of the insect's body to act upon the outside of the tympanic membrane.

If the sound is coming from the right side, it will first move the membrane of the right ear. This movement will be transmitted through the trachea and strike the inside of the membrane of the left ear. Simultaneously, sound will travel around the body to strike the outside of the left ear. If there is no change in the intensity of the sound waves within the tracheal system, and if the distance is the same inside as outside, then the internal sound and the external sound will be the same, but working in opposite directions on the tympanic membranes. It is like a push-pull system on a door. If the pushing is the same inside and outside the membrane won't move, and so the sensory cells are not excited.

This is known as a pressure difference system, a principle which has been used in directional microphones for many years. Nobody had thought to look for the same system in insects until Ken Hill revealed that this is the way it works. Differences of 25 dB between the left and right ears are achieved, even at low frequencies. Indeed, the system would be better for low frequencies than the normal pressure system of bush crickets, and even mammals.

The transmission of the sound signal through the tubular tracheal system of the insect's body will be frequency dependent. The dominant frequency in the call must match the natural resonance frequency of the tube, otherwise the push-pull interaction does not occur at the correct frequency. If the frequency changes, the relationship between the internal and external sounds changes.

At some frequencies the sounds will cancel with no excitation, while at other frequencies, the push-and-pull might be uneven and the membrane move more violently one way, with a large stimulus to excite the nerves. This imbalance also occurs when the sound source is moved around the insect. The membrane can move from a position of equilibrium to a position of maximum oscillation depending on where in the outside world the sound source is located. The distance between membranes remains constant inside but will change outside. The insect can compare the response it gets from the right ear and the left ear to locate the sound. If there is a large difference, the sound source will be to one side. If no difference, then the sound will be coming from either directly in front or directly behind.

When the sound signal reaches the nervous system and is converted into coded nerve impulses, quite separate and distinct batches of nerve fibres respond to signals of different intensities. Some fibres are not very good for low intensities, good at intermediate intensities, but inhibited at high intensities. Others may be very good at low intensities, and not so good at intermediate or high, and a third group may cover the high intensities but not the low intensities. Input from the two ears, with a particular intensity, will excite, preferentially, one of the batches of fibres.

Brian Lewis feels that this coding of intensities might be arranged in the nervous system into a space map. Particular fibres responding to different intensities might occupy distinct positions in the nervous system which would be correlated with a position in space. As yet there is no evidence that this is happening, but Lewis believes there must be some form of mapping as the behavioural response of singing insects is good.

Another intriguing problem is why, leaving aside differences in transmitting and receiving structures, crickets produce low frequencies and bush crickets sing at higher frequencies. Environment is an important influence on the sound passing from one individual to another. In general, low frequencies propagate much further than do high frequencies. There is therefore an advantage in using low frequencies to communicate over long distances. The disadvantage is that pure tones of low frequency do not give very much directional information. An insect using a low frequency sound can only identify a broad segment of space as the direction of a sound source. It must perform scanning movements to locate the transmitter more accurately. This, interestingly enough, is reflected in the listening and approaching behaviour of crickets and frogs to a sound source. They move towards the sound in a zig-zag pattern.

It is far better to use a broad band of frequencies. A greater frequency range will give directional information for a number of different frequencies, and so get better acuity of localisation. Insects have a problem. To produce a broadband signal they need to push up to the higher, ultrasonic frequencies, and lose out on distance propagation. Insects, therefore, must compromise. In bush crickets, for example, species using lower frequencies and pure tones are dispersed more widely in the environment. High frequency callers congregate.

Another interesting problem in sound direction is that faced by the female of the species when she gets close to the male. The female bush cricket, for example, knows the male is in a bush somewhere but, as she approaches the sound source, all her receptors are saturated by the high intensity sound. She loses direction information when she most needs it.

Research in West Germany has shown that the response of certain central nerve fibres, in bush crickets, to the species song were heavily influenced by vibration. They were able to record the activity of single nerve fibres in response to playback of the species song, and found that many fibres did not respond accurately. If, however, at the same time as they produced the airborne sound, they vibrated the legs at a lower frequency, at about 100 Hz, then the fibres coded the song. The central nerve fibres receive and process information, not only from the airborne receiver in the leg, but also from vibration receptors in the leg in order to code the song pattern accurately. Is there any biological significance in this low frequency vibration?

Brian Lewis, together with the West German team, looked at vibration receptors on the legs of bush crickets and discovered a group of receptors activated at 100 Hz, and responding to only 0.01 cm per second of displacement which is very slow. This was one of the most sensitive vibration receptors ever recorded. The frequency of 100 Hz was chosen because that was the syllable rate of the insect's song. So, Lewis speculated that the movements of the wings in song production produce a low frequency vibration which passes to the substrate via the legs. A male bush cricket grasps the stem of its bush during song bouts. The vibration would be propagated in the stem and detected by the approaching female. On reaching a fork she could put one leg on one branch and one leg on the other branch and compare the intensity of vibration before deciding which branch to follow. Similar observations have been made by Glen Morris, of the University of Toronto, with the neotropical katydid *Copiphora rhinoceros* in which vegetation-conducted body vibrations are combined with an 8,700 Hz airborne song.

Further work in Germany and Denmark showed that certain frequencies, indeed, do propagate through the stem, but in a very complicated fashion. The vibration travels up the stem, to be reflected from the tip and back down again. It also goes down the stem to the roots and back up again. Somewhere in the middle is a big bang as the two wave forms meet. It is not known how the insects evaluate the information, but behavioural studies have shown that sound location is greatly improved when the substrate is vibrated at the same time as airborne sounds are produced. Bush crickets, it seems, do not have the simple auditory system that was first thought. Instead they have a complicated reception system, involving a number of different types of stimuli, all analysed and coded, with the information passing into the central nervous system where a complex process of integration is going on.

Many types of orthopterans sing in chorus. Male North American gomphocerine grasshoppers *Syrbula admirablis* sing simultaneously so that bursts of deafening song activity alternate with periods of complete silence. Why insects should do this is not clear. Males may be trying to interfere with each other's songs. Also, location of individuals by predators is more difficult if the sounds are coming from different positions all the time.

Some species go even further. Snowy tree crickets *Oecanthus fultoni* synchronise their chirping in such a way that a tree-full of insects appears to pulsate with sound. Synchrony seems to reduce interference, and may even be constructed as cooperation between males to attract females to the tree. Some insects synchronise their chorus only when the population is dense. Others sing duets, individuals alternating with each other. In the katydid *Pterophylla camellifolia* alternating males sing more slowly than those singing solo.

In several species of phaneropterine katydids (bush crickets) females answer back. *Scudderia texensis* males, for example, have two courting songs – one fast, the other slow. Females respond to the slow-pulsed song by approaching silently. When they hear the fast-pulsed song they stop and answer. The male then approaches the female. When the female answers, nearby males may well attempt to intercept and mate with her. Some males, though, have developed a way of outwitting their rivals. After the female has answered, the calling male gradually reduces the call rate of the fast-pulse song, which induces the female to remain silent and approach closer, without informing other 'sneaky' males.

In the *Syrbula* grasshopper, the entire male population is aroused by an answering female. They surround the courting couple waiting to pounce every time she calls. The calling males flick their wings vigorously in order to ensure that the female knows exactly which individual should be receiving her attention.

Female mormon crickets *Anabrus simplex*, studied by Darryl Gwynne, of the University of New Mexico at Albuquerque, respond to the songs of a male by fighting with each other for the privilege of mating with him. The successful female climbs onto the male's back, and if she is heavy enough mating takes place; too light and she is turfed off. The heavy females appear to have most eggs and enough food reserves to be fitter mothers in the hostile desert environment.

Another interesting relationship was identified by William Cade of Brock University, Canada, between the North American field cricket *Gryllus integer* and females of the parasitic tachinid fly *Euphasiopteryx ochracea*. In the same way that frogs have callers and satellite males in a chorus , so too do these crickets. The males with the loudest calls attract the females, which the satellite males attempt to intercept. The callers also attract the parasitic flies which deposit larvae on the crickets. The larvae infiltrate the male crickets, eventually killing the host. The satellite males tend to escape parasitisation. There appears to be a balance of advantages to callers and non-callers, which is maintained to such an extent that two types of crickets have evolved – those that call throughout life and those that don't.

Bark Beetles

In the group of insects studied by Martin Birch, of the Hope Department of Entomology at Oxford University, it is the female and not the male that sings. The group includes the bark beetles (Scolytidae). In the engraver beetles *Ips spp.*, the male selects a tree, makes a hole and excavates a nuptial chamber. The female, attracted by scent, arrives at the tree, locates the hole, but is prevented from entering. The male has his back end sticking out and blocking the hole. When the female has determined that the male is of the right species, she starts to stridulate, rubbing a file on the back of her head against a plectrum on the inside of the thorax to produce a series of chirping sounds. The male then backs out, the female enters the nuptial chamber, and they mate. In the engraver beetles, males will mate with several females, though only with a certain number which varies from species to species. Somehow, the males can count. One species of *Ips*, for example, takes three females, while another species takes four. When an additional female arrives, she may stridulate outside the hole for a long time, but she is not allowed in. She will give up and look elsewhere.

In another genus of bark beetles, *Dendroctonus*, it is the female that attacks the tree first and the male that sits outside and stridulates. The sound producing organ in this case is at the back of the body, where a file rubs against the two wing cases. The female only mates with one male. Once a male has been accepted, both the male and female secrete pheromones which deter other males.

The Fruit Fly

A most unlikely sound producer is a minute fly – the fruit fly, *Drosophila*. It doesn't scratch, rasp, rattle or shake. When a fruit fly goes courting the male sticks one or both wings out at right angles to the body and vibrates them to produce a low frequency sound signal.

Fruit flies have a complex courtship ritual. A male will approach a female and tap her abdomen with his front legs. Through receptors on the forelegs, the male receives chemosensory information about the species identity of the female. If she belongs to the wrong species he terminates the courtship procedure and goes away to try another female. If she belongs to the right species, the male follows closely behind when she runs away, or turns towards her if she remains stationary. He then extends one wing (in the case of *Drosophila melanogaster*, the common fruit fly or vinegar fly) and vibrates it in the horizontal plane to produce a sound. Several seconds, or even minutes of sound production might pass before the female accepts the male. Before copulation the male licks the female's ovipositor. He then copulates.

In the early days of fruit fly research it was not known why male flies vibrated their wings; it could have been sound production; it might have been the wafting of pheromones – nobody was sure. Then Arthur Ewing and Aubrey Manning at Edinburgh University, and Henry Bennet Clark, now at Oxford University, managed to record these tiny insects.

They achieved this by removing the protective grill of a sensitive ribbon microphone, leaving exposed the tiny ribbon set between the poles of a magnet. The ribbon oscillates to produce the electrical signal which can be fed to an amplifier and loudspeaker, or recorded on a tape recorder. A pair of fruit flies can be introduced into a small gauze cell, fixed about a millimetre above the ribbon, and the minute vibrations can then be picked up.

The sounds are inaudible to the human ear and require considerable amplification to be detected at all. The fruit fly is only 2 mm long; the wing only 1.5 mm. The wing moves with an amplitude of about 0.1 mm, so the sound level is extremely low. To the female, though, the sound is very loud because the male gets very close to the female when vibrating his wing, perhaps only a millimetre away. Fruit flies are not interested in long distance communication. A centimetre away the female would be unable to hear the male.

Some fruit fly species use wing flapping as a visual rather than a sound signal. The Japanese species, *Drosophila suzukii*, which has little black tips to its wings, produces a kind of semaphore. In *D. melanogaster*, the Edinburgh researchers are sure the signal is acoustic. An early experiment was to remove the wings of male fruit flies so that they could no longer produce any song. These flies managed to mate eventually but instead of taking two minutes or less, they took over half-an-hour. The female required a lot more of another form of stimulation, probably chemosensory. If electronically simulated male song was played while the wingless males were courting, it only took the normal couple of minutes for mating to take place – fairly conclusive proof that wing vibration provides an acoustic stimulus.

Fruit fly songs tend to be delivered in a series of intermittent pulses. There is a pulse of wing vibration, a gap, a pulse and so on. Insects are good at beating their wings fast and continuously, but not intermittently. The thorax of a fly is

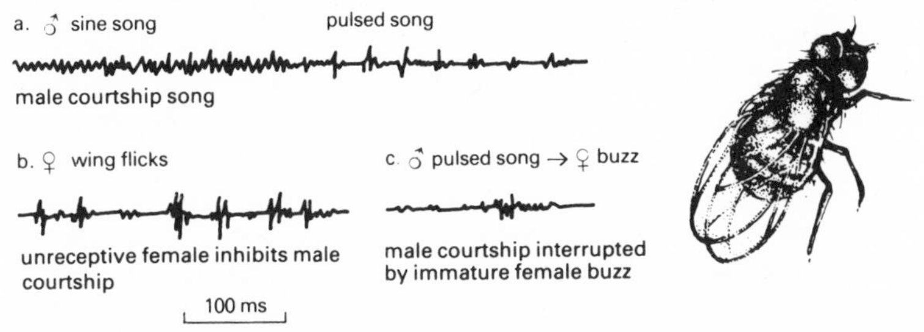

Fig. 42 Sounds produced by the fruit fly *Drosophila melanogaster*.

like a resonating oscillator; once it is started it keeps on going, and there does not appear to be a mechanism to stop it. Arthur Ewing was interested in how the fly is able to do this.

He put electrodes into the muscles of the thorax while the fly was vibrating its wings, and recorded the electrical impulses – the message from the central nervous system to the fly's flight muscles. Ewing explains his findings in terms of the way a motor car works. There is in the thorax the flight motor, the engine which drives the wings. There is also a mechanism like the clutch of a motor car, which couples the flight motor to the wings and uncouples it again. For normal flight the fly couples up the flight motors to the wings and flies off. During song it uses the clutch in a similar fashion to couple and uncouple the wing at a specific, species-determined, series of intervals.

Each species of fruit fly, and there are as many as 2,000 so far identified, probably has its own distinct song. The Edinburgh researchers have isolated and recorded about a hundred, all of which are different.

D. melanogaster has a quite complex song, in two parts. There is a continuous humming (known as the sine song because of its wave shape) and a staccato series of repetitive bursts. Ewing believes the two components serve different purposes. The continuous part is thought to stimulate the female to make her receptive. The interrupted section, which consists of pulses exactly 35 milliseconds apart, is thought to be the species identity. The inter-pulse interval is very different in closely related species. *D. simulans*, for example, has a song similar to *D. melanogaster* except that the pulses are 47 milliseconds apart.

Ewing has taken this study further and has tried to work out an evolutionary progression based on a continuum of song types. He is using as his subjects fruit flies related to *D. repleta* and is attempting to document their phylogeny.

The vast number of variations of fruit fly song is probably necessary because so many different species are likely to be found together in the same area. Around a piece of rotten fruit under a tree, there may be half a dozen species present, all courting at the same time. Although each species can recognise its own members by scent, the frenetic activity does not allow the luxury of time that olfactory senses require. A female must be able to make a snap decision about the suitability of the suitor, and song enables her to do so.

Fruit flies have a pair of feathery antennae as sound receivers. This can be proved quite simply by removing the antennae or glueing them down so that they cannot vibrate in sympathy with the sound waves of the male song. The female becomes unreceptive. Arthur Ewing has shown that the range of hearing of the female fruit fly antennae is similar to the range of the male's songs. The antennae respond to sounds as low as 100 Hz, and up to 600 Hz. Of the songs analysed so far none has fallen outside this frequency range.

In the confusion around a food source, a female must not only ensure that she is accepting the right species but also check that the male she receives in mating is the one that was doing the courting. Henry Bennet Clark, working with an African fruit fly, *Zaprionus spp.*, discovered that the male has a singing strategy to reassure the female. He continues to sing throughout copulation, and even some time afterwards. So although the female cannot see her partner, she has a sound signal which reassures her that she is still mating with the right species, and prevents sperm rejection after copulation.

In several species of fruit fly, males and females appear to sing in duet during courtship. This is a double-check system to ensure correct pairing. In *D. americana*, a duetting species, the male will not attempt to copulate until the female displays an acceptance posture. When she has stopped singing and raised both wings in a high V-shape, the male knows he can mate with her.

How, though, does a receptive female fruit fly decide what is an adequate amount of courtship on the part of the male? Does she accept on the basis of a single word or does she need to have a conversation?

There is some evidence to suggest that females memorise courtship information from previous encounters. They need to be courted for a number of minutes before they accept. The Edinburgh researchers isolated female flies and played male song of the right species to one group and song of the wrong species to another. A batch of wingless, and therefore soundless, males was released among the females. It was found that females that had been exposed to the right species song for about five minutes before the males were let in were more highly aroused and mated more readily than those exposed to the wrong species song. In a second experiment, the females were exposed to playback of the correct song for about five minutes. The tape recorder was switched off and they had two and a half minutes of silence. The males were let in and the song played again. This time the females were slower in allowing males to mate. Female fruit flies clearly need to be seduced by male song before acceptance and mating.

After courtship and mating the female eventually lays eggs, but there is a short delay as the eggs are prepared. During this egg development and laying period the female becomes sexually unreceptive. Immature females also tend to be unreceptive. When the males court, they do so with considerable vigour and harass the females into acceptance. Unreceptive females, then, must have a signal to deter over-amorous males, and indeed, they do have a song which discourages the male's courtship behaviour for a few seconds, time enough for the female to make her escape. Interestingly, the female rejection signal is the same for all species, a master-switching off signal, which also might be used to reject males of the wrong species.

So, these tiny fruit flies have surprisingly complex courtship behaviour and a remarkable song repertoire, but, as Arthur Ewing is quick to point out, most of the work has been carried out in confined spaces in the laboratory. An observation cell is only a couple of centimetres in diameter. Also, the tendency is to put a male with a virgin female to ensure rapid courtship and mating. In the wild, flies can escape more easily, and there will be fewer virgin flies about. Arthur Ewing and the Edinburgh team are now observing flies in larger, more natural arenas with mixtures of fertilised and virgin females, and experienced and inexperienced males, in order to see what is the pattern of courtship. Already they are seeing differences. Courtship in unnatural conditions tends to be continuous. A male may court for a couple of minutes, singing for about a quarter of the time. In the more natural situation there may be bouts of courtship lasting only ten or 15 seconds, of which three seconds is song. A female may sample the ten or 15 seconds, and if she's not interested fly off.

Song production in flies may not be unique to fruit flies. Other two-winged flies, Diptera, are found to do the same thing. In a greenhouse, for example, there are sometimes tiny midges whose larvae live in the soil of potted plants. If you watch them closely you can see adults running around over the surface of the soil vibrating their wings just like *Drosophila.*

Caddis Flies

One of the most unusual insect sound producers was studied by Susan Silver, working with David Pye at Queen Mary College, University of London. The insect is a caddis fly larva *Hydropsyche spp.* which lives in streams. It is not the case-building caddis but another smaller type which builds a tiny shelter of silk. In front of the shelter the larva spins a web which catches food particles in the water current. Sites for building shelters are at a premium and so each caddis fly larva defends a territory around its shelter against intruders.

When a rival enters the territory, the resident larva rushes out to challenge it. The resident lifts its forelegs, rests them just under its 'chin', and rubs pegs on the foreleg against grooves on the 'chin', producing an ultrasonic sound, which can be recorded with an ultrasonic hydrophone. Although able to identify the way the sound is produced, Susan Silver could not find how it is received. The larva has no recognisable ear. As bodies are more transparent to sound in water than in air it could be that there is a receptor somewhere inside. Alternatively, parts of the body are covered in minute hairs, and these might pick up vibrations of the water molecules and thus monitor the particle velocity of the sound.

This interested David Pye, because particle velocity is very much higher in the near-field, very close to the sound source, perhaps within a wavelength. It fades very rapidly until the sound becomes a pure pressure wave in the far-field. The caddis fly larvae are communicating at just about the near-field distance. They would be deaf, however, to sounds at the same frequency further away, such as those in a noisy stream. The caddis, thus, can hear its own sounds and not the background noise.

David Pye feels that if the interpretation put on the caddis larva behaviour is correct, then all sorts of underwater crustacea and insects, most of which are covered with hairs, could be using this form of particle velocity sound communication. This might mean that sea slaters, sand hoppers, water lice, fairy shrimps, mosquito and dragonfly larvae, diving beetles and even water fleas, could use sound to locate and communicate with each other. Even a rock pool or a freshwater pond may be a very noisy place in which to live.

Other Insects

Honey bees produce sounds, and patterns of sounds have been recorded from inside the hive. What many of them mean, and whether they are linked to particular aspects of behaviour, we do not know. Some sounds, however, have been identified. The 'old' queen bee makes a piping sound to the young queens which have not yet emerged. The young queens 'quack' back to tell her that they are ready. The piping serves to tell the young queens that the old queen has not yet left the hive. The sound also causes some of the workers to gather around the old queen prior to their leaving the hive.

Forager bees executing the 'dance', which locates distant food sources, produce a buzzing sound. The buzzing, perceived by other surrounding workers with their antennae and legs, gives distance information about the food source.

When danger threatens, a colony of alarmed bees starts to hiss. The sound is transmitted from one bee to the next until all hiss in unison. It is thought to mimic snakes and appears to succeed in frightening away large mammals, such as bears, that are a danger to the colony.

Ants have a file-and-tooth system between the thorax and abdomen. Above ground they produce sound from the file-and-tooth movements, which is radiated as airborne sound. It has a frequency around 20,000 Hz. Leaf cutting ants *Atta cephalotes* and *Acromyrmex octospinosa* stridulate when fighting with workers of neighbouring colonies. Nest mates as far as eight centimetres away can hear the sound. When buried, most of the sound energy is transmitted as substrate vibration at about 3,000 Hz. Other ants can pick up the vibration with receptors on all legs, although receptors on the fore, mid and hind leg pairs have different sensitivities. As a substrate vibration is attenuated very rapidly in the ground, an ant can identify the direction of, say, a buried companion by comparing the input from the three pairs of legs. Ants can detect, locate and extract a buried ant at a distance of about one centimetre.

Bugs also produce substrate vibrations. In Yugoslavia, researchers are looking at the little green bug, *Nezara spp.* These apparently drum on the leaf surface and the pattern of drumming is transmitted along the leaf and down the stem. Vivienne Harris, when at the University of Georgia, examined the courtship rituals of male stinkbugs *Podisus spp.* Females attract males from a distance with pheromones, and then listen to their songs to select a mate. The male sings a 'croaking' song at first and if the female appears receptive he changes to a faster 'whining' song. He approaches close to the female and butts her with his head. If she is still responsive he switches to a 'keep-away-rivals' song.

Using the scanning electron microscope, entomologist James Amrine and electrical engineer Mark Jerabek, at West Virginia University, have looked closely at the sensorium at the rear end of the flea, and identified what looks like an antenna array of hairs that might be sensitive to high frequency sounds. The sounds, they suggest, are produced through spiracles along the side of the flea's abdomen.

Many types of beetles are sound producers. The 'ticking' sound of death-watch and drugstore beetles (Anobiidae) and false powder-post beetles (Bostrichidae) is familiar. Males and females tap with the head and prothorax against the walls of their passages in the wood. It is thought the vibrations are transmitted and detected through the wood. Passalid beetles have a wide vocabulary. They live in well-organised social groups in tunnels below the bark of rotting logs. The larvae stridulate with what Michael Jarman, of the University of Bristol, describes as a 'zizzing' sound. It is made by a modified third pair of legs. Tiny pointed, horn-shaped limbs rub against ridges at the tops of the second pair of legs. A great deal of cooperation goes on in the tunnels and it is thought that the larval sounds encourage adults to help the larvae build the pupal cases. Adults make sounds by rubbing the undersides of their wings against rough patches on the abdomen. The sounds are used when disturbed, during courtship, and in aggressive situations. There are quite distinct sounds before, during and after copulation.

In longhorn beetles (Cerambycidae) the stridulatory apparatus between the abdomen and thorax is activated by flexing the body. Stag beetles (Lucanidae) have stridulating devices at the base of the femora on the hind legs. Scarab beetles (Scarabaeidae) rub the abdomen against rough areas on the coxae of the hind legs.

Burying beetles *Necrophorus spp.* were described by Charles Darwin in *The Descent of Man* (1871) as sound producers. The dorsal surface of the abdomen is rubbed against the tip of the elytra to produce a rasping sound. It was first thought that the noise accompanied the discharge of foul-smelling liquids as a defence mechanism, but Charles Lane and Miriam Rothschild, writing in the Proceedings of the Royal Entomological Society of London, were struck by the similarity of the sound to that of the bumble bee *Bombus spp.* and after investigations have concluded that the beetle is mimicking the bee as a protection against insect-eating ground predators and birds.

Wherever entomologists look, insects continue to surprise us. Mosquitoes produce meaningful sounds from their wing vibrations. The flight tone of swarms of female *Aedes aegypti* was shown, as far back as 1948, to attract male mosquitoes. To avoid confusion the flight tone of males is sufficiently different from that of females so that males cannot hear the flight tone of other males. Immature females also have a flight tone outside the males' hearing range and do not attract suitors. The flight sounds only become audible to the males when the female is sexually receptive. The sounds are received by the Johnson's organ at the base of the antenna. Flight tone, though, changes with temperature, and males are able to compensate for temperature variations between 12° and 24°C.

The death's head hawk moth *Acherontia atropos* is said to mimic the piping of the queen bee to gain admittance to the hive – whether this is to steal honey or just to stay in the dry we do not know. Tiger moths (Arctiidae) produce rasping ultrasonic sounds that confuse the radar-system of an approaching bat.

The pupae of some lepidopteran species produce sounds by scraping the body against the wall of the pupal case or by stridulation. It is thought the sounds are defensive. In one species from Central and South America, *Heliconius erato*, there is thought to be some communication between adult males and the female pupae. Sonic or olfactory signals produced by the pharate adults inside the pupal case invite males to copulate with them before they emerge. In this case there is no female choice.

Peacock butterflies *Inachis io* with their conspicuous eye-spots on the upper surface of the wings, make a startling hissing sound when the wings are flashed open to frighten off a predatory bird.

12

FISH

Jacques Cousteau, the intrepid underwater explorer, called his book *The Silent World*. 'I should have been a pair of ragged claws scuttling across the floors of silent seas', wrote T. S. Eliot in the *Love Song of J. Alfred Prufrock*. Wordsworth, in his *Memorials of a Tour in Scotland* had a cuckoo-bird, 'breaking the silence of the seas among the farthest Hebrides'. Coleridge's *Ancient Mariner* described how, 'We were the first that ever burst into that silent sea'. But the poets and writers got it wrong. The sea is not silent. Aristotle and Pliny knew that, and fishermen have, since ancient times, used pipes, oars and sticks to transmit underwater sounds to the ear, and thereby detected the presence of the best fish.

Several hundred species of fishes have been identified as sound producers, and three basic mechanisms of sound production have been described. There are stridulatory sounds which involve the rubbing together of teeth, fin spines or bones; hydrodynamic sounds caused by rapid changes in direction or speed; and swim bladder sounds produced by muscles attached to the resonating chamber.

Many of the teleost or 'bony' fish have pharyngeal denticles or teeth which can be rubbed together. Grunt fish of the family Pomadasyidae are good examples. The white grunt *Haemulon plumieri* and the margate fish *Haemulon album* are reported to use the swim bladder as a resonator to enhance the quality of the rasping grunt. The sounds are produced during feeding and also appear to mediate schooling behaviour. The croaking gaurami *Trichopsis vittatus* uses a pharyngeal rasping sound during courtship and for territorial proclamation.

The trigger fish *Balistes spp.* inadvertently produces sound during feeding, as it bites into hard coral. The fish has also been known to gnash its teeth when not feeding. On the back of the trigger fish, the anterior dorsal fin makes a low frequency vibration when it is raised and lowered. The same sound production has been recorded in sticklebacks, when the dorsal spines are moved. Sea catfish *Galeichthys felis* produce high frequency squeaks when the pectoral fin spines are flexed.

Clown fish *Amphiprion spp.*, sea horses *Hippocampus spp.*, and pipe-fish *Syngnathus spp.* rub the back of the skull against a projection on the top vertebra to make a clicking noise.

Fish make sounds as they swim. James Moulton of Bowdoin College, Maine, recorded sounds made by schools of anchovies *Anchoviella choerostoma*, and jack *Caranx spp.*, in the waters around Bermuda. When the fish were still no sounds were received. As soon as the school moved sounds, with frequencies between 500 and 1,600 Hz, were generated. If the school veered rapidly this way or that, sounds with frequencies between 800 Hz and 2,000 Hz were recorded. Whether the noise is the result of muscle and bone movements within the fish or is hydrodynamic noise outside the fish, is not clear. There is, however, some biological significance in the sound.

When the recordings of anchovies were played to yellow jacks *Caranx latus*, which prey upon anchovies, the jacks quickened their swimming rate. Moulton suggests the jacks are attracted to schools of anchovies by their swimming sounds. On the other hand, anchovies are aware of the approaching jacks.

In tests in an aquarium, sightless anchovies did not relate at all to a nearby school of normal anchovies as long as it was at rest, but as soon as a disturbance in the water made the normal fish stream and veer, they were joined immediately by the fish that couldn't see. Moulton concluded that the movements of individual fish are important in keeping a school together, and that each fish is sensitive to pressure waves created by the school, and orientates to the direction of these waves.

The swim bladder of fishes seems to be involved in sound production in a variety of ways. Basically it behaves as a resonating chamber, changing the quality and quantity of the emitted sounds. The trigger fish *Balistes spp.* also produces a sound by rubbing together bones of the pectoral girdle at the anterior end of the swim bladder. In other trigger fish species e.g. *Rhinocanthus*, the pectoral fins are rubbed against the sides of the body where the swim bladder is covered only by a thin layer of skin. *Balistes* has been reported to bang its pectoral fins on the side of the body next to the swim bladder.

Some bony fish, the physostomes, have the swim bladder connected to the oesophagus or stomach by a tube. They include the bony tongues (osteoglossoids), the mormyrid fishes, some of the salmon species and pike (Esocidae), and the herring-like fishes (Clupeidae) and eels (anguilloids). As a physostome fish rises in the water, air is expelled from the swim bladder and bubbles come out of the mouth or gill chambers. Each bubble released is accompanied by a squeak. Fishermen were able to spot a rising shoal of herring by the minute bubble disturbances on the surface of the water. The noise under the sea must be deafening!

Perhaps the noisiest of fish, and the best known for the sounds they make, are the drum and croaker family (Sciaenidae). Large sonic muscles, derived from the body wall, lie alongside the swim bladder. Contractions of the muscles produce short bursts of drumming sounds. The muscles are only present in males.

Catfish (Silceroidae) have a modification of the 'drum and croaker' configuration. A thin springy piece of flat bone, the *Springfederapparat* or 'elastic-spring', lies over the anterior dorsal part of the swim bladder. Contraction and relaxation of sonic muscles causes the spring to flex and relax, which in turn vibrates the swim bladder.

In some fish, notably the toadfish *Opsanus spp.*, the sonic muscles are attached directly to the heart-shaped swim bladder. The fundamental frequency of the sound produced is exactly equivalent to the frequency of vibration of the sonic muscles. The 'horn-like' or 'boat-whistle' sound of the toadfish *Opsanus tan* consists of a short 240–300 Hz grunt followed by a 250 Hz pure tone burst. It is not known whether both sexes produce the sounds, but they are heard usually prior to the start of the breeding season.

Anthony Hawkins, at Aberdeen's Marine Laboratory of the Department of Agriculture and Fisheries for Scotland, has been studying the sound production and hearing abilities of commercially important fishes in the North Sea. Haddock *Melanogrammus aeglefinus* produce calls which consist of single or repeated knocks or grunts. Knocks are usually of two closely spaced syllables, while grunts may be made of up to 16 pulses. Knocks, delivered at different rates, tend to be related to various levels of excitation, particularly in aggressive encounters. Grunts are heard more frequently in defensive contexts, for example in the escape sound after a fight. A fish like the haddock in the spawning season is producing sounds almost continuously. The sounds appear to be used both in fighting other males and in courting females. A hydrophone dropped into the North Sea during February picks up a cacophony of haddock love-calls.

It is not wholly clear in what context a particular sound is used. Hawkins has found, however, that under certain circumstances a fish will change the rate of its calls. A haddock which is mildly aggressive to another will produce a low rate of calls. As it becomes more aggressive the knocks are delivered at such a fast rate they sound like a hum. Rival males are chased and rammed.

Very little was known about the breeding behaviour of haddock until some were kept in a 700 gallon, glass-fronted aquarium. Before the females were ready to spawn the male fish would carry out their aggressive behaviour, facing and then swimming broadside to each other with their fins spread out in full display. In the tank one male became dominant and it was this fish that mated with the single female. The female remained silent, swimming near the floor of the tank. The dominant male swam around her in a tight circle and the knocking sounds he produced became gradually faster until the sound was a continuous purring, much like a motorbike starting up and moving off. At the same time a marked change occurred in the appearance of the male. Dark blotches could clearly be seen along the sides of the body and the fins turned a darker colour. As the female spawned the male became silent and the pair

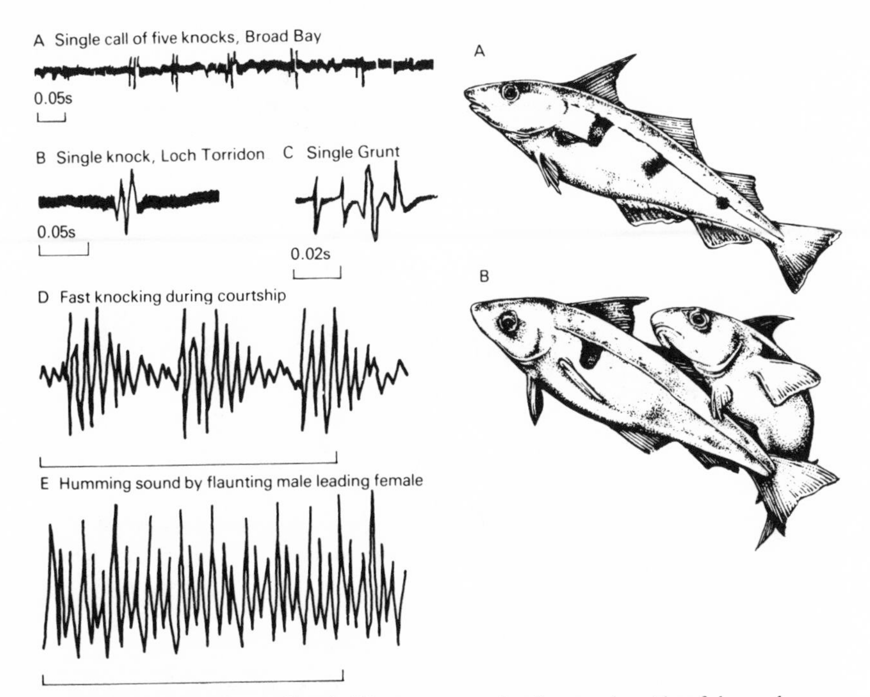

Fig. 43 Male haddock calls. A. Pigment spots develop on the side of the male haddock when displaying to the female. B. Male haddock 'embraces' the female.

embraced, swimming upwards. The female laid about 10,000 eggs and the male shed his milt to fertilise them. It is thought that the sound made by the male before spawning induces the female to spawn.

Cod *Gadus morhua* produce deep grunting sounds, either as single grunts or in groups of four grunts. V. M. Bawn found that grunts are emitted during defensive and aggressive behaviour, and by both sexes, although males tend to use grunting in the courtship of the female.

The sounds made by lythe *Pollachius pollachius*, have been recorded by C. J. Chapman, of the Marine Laboratory, Aberdeen. Juvenile fish, common in inshore waters from May to October, characteristically gave repeated bursts of short grunts during competitive feeding and aggressive encounters. Breeding behaviour of adults has not been studied.

The heavy, bottom-living tadpole fish *Raniceps raninus* give single short grunts when alarmed, but little else is known of their sound repertoire.

Anthony Hawkins, together with A. D. F. Johnstone, looked at the hearing of the Atlantic salmon *Salmo salar*. Fish were trained in the laboratory to respond to sounds, and it was found that salmon only respond to low frequency tones, below 380 Hz. It was concluded that sound is not as important for salmon as it is for other fish. Its ability to distinguish signal from noise is poor. Russian workers, though, have shown that salmon emit 100–500 Hz

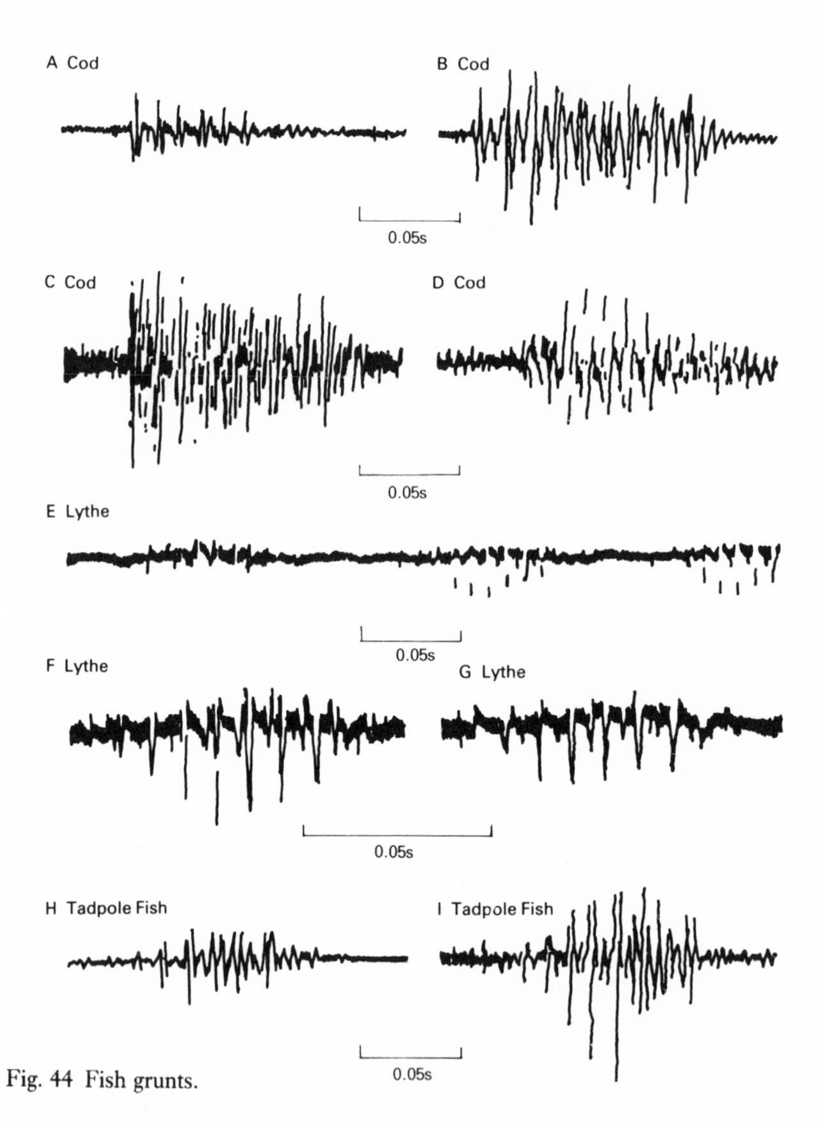

Fig. 44 Fish grunts.

sounds, which are thought to be given during the spawning period.

Hearing could be important for salmon parr as a number of aquatic prey species produce locatable sounds. Water boatmen *Corixa spp.*, and *Notonecta spp.*, and the stoneflies *Dinoceras spp.* and *Chloroperla spp.*, are known to emit sounds. Equally, hearing could have some survival value for the salmon itself. Any aquatic mammal, such as an otter or seal, would generate sounds as it breaks the surface in order to breathe.

As for the angler: a quiet conversation on the riverbank is unlikely to disturb your trout or salmon, but a careless footfall will generate sufficient water-borne sound to be detected and have all fish in the area scattering for cover.

13

ROARS IN THE DAY, SCREAMS IN THE NIGHT

The Roaring of the Red Deer

Autumn on the Isle of Rhum, off the west coast of Scotland, is the time of the rut. Dominant red deer stags greet all challengers with loud, bellowing roars. These vocal conflicts last for most of October. It is the breeding season of the red deer *Cervus elaphus*, when the stags spend a considerable amount of time and energy competing for the hinds. Standing on the sidelines have been researchers from the University of Cambridge. Tim Clutton-Brock, of King's College Research Centre, in particular, has been concerned with why red deer roar.

Many animals have dangerous weapons. Deer have antlers, rhinos have horns, and elephants have tusks, but strangely, the animals rarely use them. Animals are often in conflict with competition for mates or access to food, but they hardly ever use their sophisticated weaponry. Physical confrontation is costly. A rival is likely to have the same array of dangerous weapons and little can be gained by such an encounter. Even if the rival is beaten, there is a fair chance of damage to the victor as well as the loser, and a damaged animal may not be able to mate or use the food that it has just won.

During the course of a three week rut stags, on average, fight five times with rivals. Although stags do not kill each other, some do receive injuries. Of the 50 stags in the Rhum study area, about 5% receive permanent injuries, like a broken leg or an eye gouged out. Stags can be expected to rut for five seasons, and so there is a 30% chance of an individual being maimed. Clearly, it pays a member of a species with dangerous weapons to avoid using them, wherever possible.

How, then, do individuals confronted with a rival settle a difference of opinion about which one should have access to the females or food? There must be a reliable, and safe, method of assessing rivals, some sort of contest which would enable individuals to decide who would be the winner of a fight without actually staging the fight itself. Tim Clutton-Brock feels that this is the role of roaring contests. Stags may roar up to three times a minute during the rut. They do not feed, but spend the three weeks bellowing, holding hinds, and defending the harem from other stags. Their body weight declines gradually, until they must stop rutting. With this kind of energy consumption, Clutton-Brock felt that the roaring must be serving an important function.

In order to start to understand roaring, the research teams looked at the circumstances in which stags roared. When two harems get close to each other, the harem holder of one will often approach the harem holder of the other. One stag approaches and roars. The defender replies. The two may roar at each other for up to an hour, the roaring rate gradually increasing. Nearly always, one stag withdraws.

If the challenger fails to retreat and keeps on coming, the two stags go into parallel walking. This involves the animals in strutting up and down, five metres apart, at right angles to the line of approach of the challenger. Parallel walking may last from 30 seconds to five minutes. If the challenger persists, then a fight takes place. So, before a fight is allowed to occur there is a series of displays which get gradually more intense. Were the contestants, asked Clutton-Brock, roaring at each other to see which could roar the most? Are they roaring simply because they feel aggressive or are they trading roar for roar in a finely tuned system for assessing the physical strength of the rival?

The researchers taped the roaring of stags for playback experiments. It turned out that the animals were particularly fussy and would only respond to the very best hi-fi equipment. Carting around a very large cabinet loud-speaker, complete with amplifier and car batteries, and keeping it all dry was a major performance in itself. Nevertheless, roars were played back to stags and their responses monitored. Stags would turn to face the speaker and roar back at it at the same time increasing their rate of roaring. So far, so good. What the research team wanted to know next was, did the stags increase the rate of their own roaring to match that of their challengers?

In the next series of playback experiments, the recorded tapes were edited to give three different rates of roaring, each rate gradually increased to give a graded series. The first series of calls was at a relatively low rate giving two and a half roars per minute (RPM). Then came a period of silence in which the animals could rest, followed by a set of roars at five RPM. After another five minutes' interlude, the third series was played at ten RPM.

The rate of two and a half RPM was a level which the researchers considered even the weakest stags could out-roar. Most healthy stags should be able to beat five RPM, but ten RPM was a rate that could only be kept up for five minutes. Six stags were presented with the tapes. They easily outroared the two and a half RPM playback, worked a little harder on the five RPM, but at ten RPM they tended to stop roaring and in some cases they even rounded up the harem and made a rapid exit.

The next question was whether a stag's roaring ability is related to his fighting ability. There is no point in giving a frightening display if it cannot be backed up. To check this out, the researchers looked at how differences in

Fig. 45 A red deer stag dispute: A. Roaring as a means of assessment. B. Parallel walking. C. Stags engage and lock antlers. D. Victor emerges and chases rival. (After Tim Clutton-Brock)

fighting ability among stags, which were known from data collected by Fiona Guinness and Robert Gibson during the previous five years, related to differences in roaring rates in the playback contests. It turned out that the stags which were the best fighters were also able to give the highest roaring rates.

Clutton-Brock suggests that roaring is tiring, and that during the process the animal gradually becomes deoxygenated, and the rate of deoxygenation is related to the condition of the body, physical fitness and strength. Fighting ability is also related to physical fitness and strength. The rate, therefore, at which a stag can roar should give an accurate and reliable assessment of how the animal might perform in a fight.

That, though, is not the end of the story, for the picture is more complicated. Clutton-Brock looked at other aspects of roaring. He found, for example, that there are differences in pitch between stags. Young stags, below adult body weight and size, have a significantly thinner and higher pitched voice. Mature stags can distinguish the playbacks of immature roars. Clutton-Brock feels that there are qualitative, as well as quantitative differences in roars, that are related also to fighting ability.

Tracing back a stag's records, to the time it is born, the researchers think that an individual's fighting ability might be related to its growth during the first two years of life. In fact, there is some evidence that the size and strength of a young stag is linked to the amount of milk it gets from the mother, and that roaring is related to rearing.

The Red Fox

At the beginning of the year, throughout Britain, the silence of the night is often shattered by the startling scream of the red fox. In the country, and nowadays very often in the city, both dog foxes and vixens call to each other with a one or two syllable shout. It is known as the vixen's shriek, and on a still, quiet, frosty night there is nothing more eerie. To David Macdonald of Oxford University, the call is evocative of the mystery which surrounds it. What are the messages passing back and forth between these animals of the night?

Foxes are essentially solitary animals, although a dog fox and several vixens may share the same territory. The repertoire of sounds passing between them is considerable. The vixen's shriek is a piercing long-distance shout designed to communicate with other foxes over a large range. There are other calls, more subtle and sensitive, which vary from moment to moment, and which might only be audible over a distance of a few centimetres. The same calls may be shouted or whispered in a series of vocalisations graded by loudness.

The vixen's shriek, also given by dog foxes, is heard mostly in January and February. This is the period of the year when foxes are known to be mating. Itinerant young males, without territories, travel the countryside looking for resident males to depose. Alpha males are on the move too. They trespass into their neighbours' territories, attempting to solicit the affection of neighbouring vixens. The entire fox population is stirred up and active.

In the confusion, David Macdonald has noticed that foxes appear to recognise individuals. He has heard one fox shrieking in one part of a wood and another answering on the other side, although, by telemetric tracking, Macdonald knew that many animals were scampering around but remaining silent. This implied that specific signals were directed at particular receivers who responded appropriately. Foxes may be able to identify each other on the basis of voice alone. Countrymen can identify what seem to be separate calls; some are high-pitched, others lower pitched. Are these different calls from the same fox or calls from several different foxes?

With the help of Cotswolds sound recordist, Ray Goodwin, who has one of the best collections of fox recordings, David Macdonald related particular calls to the places where they were recorded. Ray Goodwin felt that he could recognise individual fox voices. He would often hear a particular voice in a given place, and completely different voices associated with quite separate sites, probably different territories. Looking at sonograms of Goodwin's fox recordings, Macdonald was able to identify what seemed to be quite distinct voices. In one area, for example, a fox would give a stuttering, two-syllable bark with peaks of sound energy at 70–80 Hz and 120 Hz. A similar two-syllable fox recorded regularly in a quite separate area, showed the same two energy peaks but about 40 Hz higher in pitch. The voices appeared to be from separate individuals, but despite a decade of study researchers are still not certain that these results are interpreted in the correct way. So, the vixen's shriek still remains a mystery. Other fox calls, though, are slightly easier to understand.

The 'wa-wa-wah' call, a polysyllabic, staccato noise, is heard often during the winter breeding period, and occasionally throughout the year. Sometimes this long-distance call is greeted with silence; at other times there is a 'wa-wa-wah' reply. Using night-vision equipment, Macdonald has observed foxes making these calls when on the move. Every so often a fox moving across a field or through a wood will stop, bark, and then continue. Two foxes barking seem to move gradually closer together.

Macdonald was once perched in a tree, over a fox earth, on an icy-cold winter's night. In the distance he heard a 'wa-wa-wah' call. From another direction he heard a reply. The two individuals duetted back and forth, and gradually came closer together until they met beneath Macdonald's tree. As they approached the wa-wah's became much quieter, and changed to clucking sounds. This led Macdonald to believe that the wa-wa-wah sound is a contact call. It may say 'Here I am', or 'Where are you?', and maybe 'I am Fox so-and-so'.

Then there is 'kekkering', a call used more in close-quarters encounters. It is a harsh, staccato noise, much like a wooden ruler being rasped along a paling fence. It is heard at greatest intensity during fights. As the fight subsides, the sound smoothes out into a whine, which might be associated with submission. When a submissive fox approaches a dominant animal, it lowers its body towards the ground, flattens its ears against the head, the tail lashes and the mouth is opened with the lips pulled back a little. Accompanying the visual signals is the whining sound. Macdonald believes there is a graded continuum with a rasping, aggressive element at one extreme, and a whining, submissive element at the other. An animal will call with a different mixture of confidence and lack of confidence, depending on its standing in fox society.

Foxes have alarm calls. A vixen outside an earth with cubs, hearing or seeing potential danger approach, will give a single, staccato, projected bark. The cubs immediately fall silent and are gone in an instant.

Rats and Mice

In the city, foxes may find themselves competing at the dustbin with rats. Rats, it turns out, are very noisy animals, but humans cannot hear much of their squabblings at the bottom of the garden for they communicate mostly with very high frequncy sounds – ultrasound.

In the 1960s, one of the pioneers of rodent vocalisations, the Belgian researcher Ellian Noirot, studied the maternal behaviour of mice and made the curious discovery that virgin females and males would show maternal behaviour towards pups. They would retrieve, lick and cover them in the nursing position; behaviour normally associated with mothers. It was more marked with one- or two-day old pups than with 12- to 13-day pups. Something coming from the pup, Noirot thought, must be inducing the animals to respond in this way. She eliminated visual and tactile cues.

Noirot recalled that in 1956 the German researchers Zippelius and Schleidt had shown that baby mice and voles emit ultrasound when removed from the nest up until day 13 when their eyes open. At that point the ultrasound stops. This correlated nicely with the mice study, so Noirot suggested that ultrasound from the pups was eliciting this strange and compulsive maternal behaviour. At this stage the sounds had not been recorded, but in 1969 she found David Pye and his bat detector and was able to show that ultrasound really is produced by baby rodents.

Looking for a PhD subject at that time was Gillian Sales, now at King's College, of the University of London. She became interested in the ultrasound phenomenon in rodents and started by recording the calls of rats, mice, voles, squirrels, hedgehogs and a host of other small mammals. It was only in the myamorph group of rodents – the rats, mice, gerbils and hamsters – that ultrasonic calls were detected. Sales made more detailed studies of the intensity and frequency patterns of the calls. Animals were recorded daily and their sounds analysed to see how patterns changed with age. Until then it was thought that animals stopped calling when their eyes opened. Sales, though, found that older animals would emit ultrasonic calls under certain circumstances. She noticed, for example, that adult rats would give very high frequency squeaks when flipped onto their backs, and this happens to be the position taken up by a subordinate rat when it has been defeated in a fight. Sales, therefore, looked and listened to fighting behaviour. Lo and behold, rats produce a great deal of ultrasound, particularly when one rat attacks another, and it turns out that there are two categories of calls – 'submissive' and 'aggressive'. A short 50,000 Hz call is associated with aggression, and an up to three seconds long, 22,000 Hz call accompanies submission.

When defeated, a rat falls on its back, puts its limbs out stiffly and shows a characteristic pattern of respiration. There are very long exhalations separated by short, sharp intakes of breath. Submissive calls follow the respiratory pattern. There is some evidence that the long submissive calls inhibit aggression in the dominant animal. When two animals fight, the aggressive behaviour ceases as soon as a long call is heard. If there are no long calls the fight continues. Fighting behaviour seems to occur in fits and starts, mediated by sound signals.

When rats mate other calls are heard. The male chases the female and produces a call with two phrases. A frequency modulated portion, changing in pitch, up and down, through an octave at about 50 times a second, is followed by a sequence of constant frequency. The warble and straight tone are alternated. Although usually produced by the male, the female will occasionally emit the same sounds. The function of the call appears to be to induce the female to mate more readily. Ron Barfield and L. A. Geyer carried out a series of experiments where they presented female rats with the ultrasonic calls of males before allowing them access to castrated males. In front of a normal male a female will dart about, wiggle her ears very rapidly and hop. She will show this behaviour more to castrated males accompanied by sound recording than to those without. It is thought that the call encourages the female to take up the mating posture, suppress her aggression and so allow mating to go ahead. In further studies, Gillian Sales has noticed that if the female is very receptive the male produces fewer calls. On the other hand, a totally unreceptive female will cause the frustrated male to shout ultrasonically long and hard while chasing the female about.

Researchers have looked for correlations between the state of the oestrus cycle of the female and ultrasound emission by the male. Odours produced by the female elicit calling from the male. Vaginal swabs from the female will stimulate males to call, even in the absence of the female herself. Gillian Sales feels that this is a part of a complex chain of events that has to take place before normal behaviour patterns can continue. It is a fail-safe system involving different senses. If one fails, communication ceases.

After ejaculation, the male, then incapable of further mating for a few minutes, changes his tune and emits the 22,000 Hz submissive call. The call appears to be associated with any situation where a rat is about to withdraw from the action.

In hamsters, calling during courtship and mating is confined to females on heat. Female hamsters are normally very aggressive. They live alone and would need to advertise that they are ready for mating. The presence of a male hamster enhances calling behaviour; indeed, both male and female call back and forth to each other.

In all of the myamorph rodents, babies produce ultrasonic calls which spark off searching behaviour in the mother. A tape recording of baby calls will cause a mother to leave the nest and investigate the sound source. In a two-chamber choice test, mothers will always go to the chamber with the baby calls being played. They can also distinguish between artificial and real ultrasonic calls. Jane Smith, when at Queen Mary College, London, found that mother mice would respond to the calls of youngsters only if the mother had had prior exposure. Mothers need a bit of experience. In addition, the calls only appear to arouse the searching response. An olfactory cue is needed before the young can be found.

The baby call does not appear to be specific to a particular individual. Although there are variations in calls in a litter, there are also considerable variations in the calls of an individual. It is hard to identify species from species, let alone one pup from another. It seems to be a generalised signal.

In baby rodents the production of ultrasound appears to be related to distress and linked, in part, to the ability to regulate body temperature. New-born babies are not able to maintain their own body temperature.They simply take on the temperature of their surroundings. They are, however, peculiarly resistant. A one-day old mouse can be cooled to 4°C with little or no effect, and it produces very little ultrasound. At about six to seven days, baby mice begin to regulate their body temperature, and it is at this time that they produce the greatest intensity of ultrasound. In the laboratory, if a mother is taken from a nest, the babies will produce very little ultrasound. If they are cooled they gradually emit more calls.

When the youngsters get to about 21 days and are fully covered with hair, they can fully maintain body temperature, even in the fridge. At this time ultrasonic calling falls off. There are, of course, exceptions. One large Australian rat has its babies born hairy, and these youngsters do not produce ultrasound. But the baby spiny mouse *Acomys cahirinus* has a good covering of hair, is able to move within a few hours of birth, and does produce ultrasound.

There are also associations between calling and feeding. When the mother returns to the nest the youngsters call to trigger feeding behaviour. Work in Israel has shown that replayed ultrasound helps the milk-let-down reflex. It appears to influence the production of prolactin which helps the secretion of milk. Ray France has found that some rat mothers produce the 22,000 Hz submissive call while suckling. Nobody knows why.

Woodmice *Apodemus sylvaticus* emit ultrasonic squeaks when exploring familiar territory. Dominant animals produce more calls than subordinates. It is thought the calls might have a territorial proclamation function. Calls are produced more readily in the animal's own environment than in an area with a foreign smell. Ultrasonic calling in rodents is influenced by other stimuli such as touch and smell. The odour of another individual will inhibit the exploratory calls of adult woodmice. This makes sense, for if an animal is in another's territory it would be unwise to shout about it. In a related species, the yellow necked mouse *Apodemus flavicollis* the calls are produced only when animals get together. They often live in trees and when two individuals meet on a branch, and need to pass, they call. This is a kind of contact call. Curiously, ultrasonic calling in two closely-related species has quite different functions. In the woodmouse it appears to be saying 'keep away', while in the yellow necked mouse the message is 'I'm a friend, it's OK to pass'. Gillian Sales has found that the two mice have calls pitched slightly differently – the woodmouse at 70,000 Hz and the yellow necked at 45,000 Hz. The woodmouse has been found to have a sharp sensitivity peak at 70,000 Hz, and is much less sensitive to 45,000 Hz. The yellow necked mouse, at the time of writing, has not yet been investigated but Gillian Sales feels it would be a very nice correlation if it had a sensitivity peak at 45,000 and nil at 70,000, particularly as the two animals share the same range in Britain.

Other small mammals, such as shrews, are thought to use their ultrasonic calls for echolocation. E. R. Buchler of the USA has tested shrews by putting them on a circular platform in the dark and making them choose a step down in order to run to a tunnel in which they find a reward. He has recorded very short 40,000 Hz calls, which are produced as the animals search for the correct step down. When the animals' ears are plugged, they locate the step far less easily than when their ears are free.

14

PRIMATE TERRITORIES

Indris and Howlers

The indri *Indri indri* of Madagascar is the largest of the lemurs. The name indri is not, surprisingly, the local name for the animal – in some areas it is known as babaroto, meaning boy, or in other areas, amboumala, meaning forest dog. The name came when Sonnerat, the first European zoologist to see it, was exploring the island. His Malagasy guide pointed to a tree and shouted 'Indri, Indri', which actually means 'Look there'. Sonnerat thought it was the name of the beast and scientists have used it ever since.

One of the scientists who has been studying the social behaviour of indris and other lemurs is Alison Jolly, of Rockefeller University, New York. She found that they were so hard to find in the forest that the only way you knew they were there was by the call.

The indri's wail is one of the loudest sounds in the Malagasy forests. It can be heard three kilometres away. Indris are monogamous. There are usually just a pair in each territory, although they are often with youngsters. Females start to sing at dawn and are joined by the males shortly afterwards. In all lemurs studied so far the female is dominant. If there are juveniles present they will also join the singing after the adults. Sitting in the middle of the forest in the morning, Alison Jolly was able to hear the chorus of wails as every indri checked out the whereabouts and existence of the other pairs. Indris have strict territories but the calling system works so well that researchers have found it difficult to establish whether territories are being defended during bouts of singing. Individuals sit in the middle of their territories, answering each other, while languidly feeding on leaves.

Occasionally there are border encounters and disputes. Jonathan Pollock has observed these. The female indri takes a back seat in these events, although both pairs yell loudly at each other across the territorial boundary. The males will leap about and yell, sometimes stopping to scent-mark with their chins. After about ten minutes the two groups separate and return to the centres of their respective territories with no further confrontation.

Fig. 46 The Indris of Madagascar.

It is difficult for observers to tell whether an animal is using this kind of wailing song to defend a piece of territory or simply to keep some distance from its neighbours. Here playback experiments are important. It was film-maker David Attenborough while on an expedition in Madagascar, who carried out the first playback of singing with indris. The animals would not come out to be filmed, although the film crew could hear them all over the mountainside, so they decided to record some indri calls and play them back in order to try to entice some animals to perform in front of the camera. Sure enough, a family group of three indris materialised out of the bush, and began, not to sing back, but to bellow in fright – a kind of in-and-out sawing roar. What the film crew had done, in fact, was a critical experiment. If an animal approaches a loudspeaker playing its song, and challenges it, then it is very likely defending a piece of territory. Indris, thus, were shown to be territory holders.

In the New World, the howler monkeys are the largest of South American monkeys. On waking at five or six o'clock in the morning, a troop of howler monkeys will start to howl. The males have a low resonant howl and the females 'bark'. The chorus of bellows which is answered from troop to troop echoes around the forest. The loudness of the sound is due to modified bones around the throat which form a resonating chamber. This gives the howler monkey its characteristic head and neck shape. The male, with its larger resonator, has a thick neck with what looks like an enormous double chin.

David Chivers of Cambridge University has studied the howler monkeys of Barro Colorado Island, Panama. He found that the calls served the function of spacing out groups of monkeys. Previously, researchers had argued that howler monkeys were territorial, and that social groups defended chunks of habitat; but no-one had actually followed a troop for a long enough period of time to be sure of this. When Chivers came to following a troop for three months, it quickly became evident that howlers were not defending fixed territories with their calls. During that period, over 60% of the troop's home range was shared with neighbouring groups of monkeys. What the monkeys were doing was to keep part of their range exclusive to themselves at any one time.

By sitting in a clearing in the forest each morning, Chivers was able to plot the identity and position of groups calling, and to record how much they called. The troop nearest his vantage point were followed during the rest of the day. He found that the closer two groups had been during the morning the greater their tendency to move apart later in the day. In this way the different groups of howlers spaced themselves out very effectively.

Occasionally, two groups would find themselves very close to each other when they started to call; perhaps they had approached each other from either side of a ridge. On these occasions the two rival groups would have a vocal battle, with maybe a little chasing and scrapping, and then move apart, going their separate ways. Howler monkeys, unlike indris, are tolerant of neighbours sharing parts of their home range. In Africa, similar mutual avoidance is demonstrated by adjacent troops of mangabeys *Cercocebus spp*. A characteristic 'whoop-gobble' is given to space groups of animals.

Another group of New World primates, which has been studied by John Robinson, is the titi monkeys *Callicebus spp*., which show different territorial strategies in closely related species. The dusky titi *Callicebus moloch* has one of the most bird-like songs of all the primates. As with the indri, the song is a duet between a mated pair, which seems to be given at territorial boundaries. These little monkeys sleep in the middle of their territory at night. Early in the morning they begin to move towards the boundaries, where by eight o'clock they are ready to sing. When they hear a neighbour approaching and singing, they move towards it and sing back. The duet is so closely integrated that it is difficult to tell that two animals are calling. Females seem to challenge females more than they approach males. Males challenge males. Each pair thus cooperates to keep its own personal rivals out of the area. After the song bout is finished and they are satisfied that no other pairs are about to challenge, the pair returns to the middle of the territory for the remainder of the day.

John Robinson tried some playback experiments and got some odd results. If he played back the duet of a rival pair in the middle of a residents' territory, the residents would react far less than if it were played back at the edge. Presumably the sudden appearance of rivals in the middle of the territory was totally outside the animals' experience and they failed to respond. At the edge, playbacks enticed resident pairs to rush towards the sound and sing back as hard as they could. Robinson was even able to identify the roles played by the two animals in the pair. If he played female calls or a duet led by the female, the resident female would approach the playback speaker, stare at it, and sometimes pick it up and shake it violently. If male calls were played then the male of the pair reacted, but not as strongly as the female. Females, it is thought, have more of a stake in keeping a territory and raising their young.

The other species Robinson studied was the widow titi *Callicebus torquatus*. This monkey has a huge home range which it would be impossible to defend. When Robinson played back widow titi calls the other widow titi nearby simply cleared out of that part of the forest. The widow titi is not territorial like its dusky cousin. The widow titi song is used for spacing. Robinson found that the difference in behaviour of the two species could be brought down to a question of economics.

Dusky titis live in food-rich woodlands. Widow titis live on the almost sterile white sands of the Amazon where there is little nutrient in the trees. Widow titis unlike their richer relative the dusky titis must range very widely in order to get enough to eat. The range is too big to defend, so they simply do not try. Their songs keep them spaced away from the neighbours.

A territory can mean a number of different things. It can be used to mean a personal space around a wandering individual. It can be a nesting territory with a set distance around a nest site defended, as is the case with some birds. Then there are feeding territories, typical of primates. These are areas in which a troupe or a pair of animals obtain all the resources they need for the year. Some primates defend a feeding territory, others have 'homes'.

There is also a close correlation between duets, monogamy, and territoriality. A duet is a male and female singing in very close synchrony – the timing may be down to small fractions of a second. This close liaison between male and female would reinforce the pair bond. It would also tell neighbours that the residents are working in unison and that they would be a difficult pair to challenge. Monogamous animals need to defend a territory for the young. Monogamy tends to go with a home range with food supplies scattered fairly evenly throughout. The pattern of defending a mosaic of territories, in the way the indri does, goes with the raising of small numbers of offspring when both parents must put their energies into raising those children. Titi males and siamang males, for example, carry the babies. In other monogamous primates the male's role is to defend the babies, which he knows are his because he has his wife to himself. Monogamy and territoriality are reflected in song, not by the pair singing the same song, but by an integrated combination that is a duet. The female plays a major role in singing. Unlike some song birds, the male is not simply announcing he is a male and ready for mating. He contributes to the combined effort. The great call of the gibbon and the wail of the indri are predominantly female calls. She starts the ball rolling while the male joins in after she has started. The two animals together, though, give more information to a listener than if there was just one calling.

In a large group of animals, the variety of sounds they make, when combined, give even more information and meaning to a listener. A troup of ring-tailed lemurs *Lemur catta* make quite a din when on the move. They click and grunt, and the more excited ones miaow like cats. They crash branches as they bounce on and off, producing a variety of non-verbal sounds. Alison Jolly feels that the intensity of activity and the emotions of the group can be sensed from their collective chorus of noise. An individual, she suggests, could judge its compatriots' state of mind just by listening to the sounds being made around it. Mobbing is an example of this. When ring-tailed lemurs see an enemy on the ground they are relatively safe and tend to approach and yap like a pack of small dogs. One lemur will start to yap and the others will join in so that it sounds like one animal yapping, such is the synchrony of the sounds. The more frightened or upset they are, the louder the yapping. The loud yap also tells other troops in earshot that there is menace on the ground and it might be coming their way. The emotional level of the entire troop is announced by the combinations of sounds, not by the single stereotyped call with the single stereotyped meaning.

Ring-tailed lemurs have one of the largest repertoires of sound and gesture of any birds or mammals studied. They communicate through a lot of channels. They promenade along the ground with their black-and-white striped tails in a question mark over the back. They have a well developed olfactory sense. Individuals will threaten each other by perfuming their tails, arching them over the back and wafting 'evil-smelling' odours in the course of 'stink fights'. They mark branches with scents.

Ring-tails have a whole battery of different sounds. There is the mobbing yap, and the 'cat miaow' which is graded between a faint mew and a loud cat-like screech. Male ring-tails howl like coyotes. Animals sitting closely together purr.

Gibbons

Gibbons are the smallest of the apes and the most agile of mammals. They live high up in the rain-forest where they swing by their arms from branch to branch. Some researchers consider them as honorary birds, for not only do they sing like birds but their acrobatics leave you believing that they have wings. Gibbons have been known to jump from one tree to another 15 metres away.

Gibbons live in distinct territories in the tropical forests of south-east Asia. In the dense foliage, calling is important as a means of communication between partners as well as between neighbours and other individuals. David Chivers, together with Elliot Haimoff and others, of Cambridge University, have studied the calling behaviour of gibbons.

A special feature of gibbon songs is that they have all the sound energy compressed into a narrow frequency band. This is quite unlike humans and other higher primates, which have their calls in a wider frequency range across the sound spectrum. Gibbons produce sounds which are highly specialised for their dense rain-forest environment. The compressed signal is able to travel further, sometimes up to three kilometres, than it otherwise would if composed of wider frequency bands.

This long-distance signalling is further enhanced by the time of day at which gibbons call. Like the indri and the howler monkeys, most species of gibbons call at or around dawn. The tropical rain-forest is a dense, closed canopy with a height of about 30 metres. When the sun begins to rise over the horizon the layer of air above the canopy becomes heated, and is slightly warmer than the air below the canopy, which is still cool. The animals climb up into the tallest trees and call. Up there, the sounds are not absorbed by the ground litter. In addition, the sounds that filter through the canopy are reflected back down by the warm layer of air, so travelling a longer distance. As there is only a short period when this phenomenon occurs many species find themselves calling at the same time. There is, however, some staggering of the species in the dawn chorus; the banded langurs *Presbytis melalophos*, for instance, call before dawn and after dusk. The peak of calling of the dusky langur *P. obscura* is around dawn. Just after dawn the lar or white-handed gibbon *Hylobates lar* reaches its peak of calling, and the siamang *H. syndactylus* calls about an hour later.

The siamang is the largest of the gibbons and has the most complex of calls. It gets its message through, not by calling at a time when atmospheric conditions are right for long-distance sound propagation, but with the help of resonating vocal throat pouches. Its voice is a deep boom, but the siamang and its other, smaller, gibbon relatives do not sing alone. Like indri, male and female gibbons sing in duet. Each gibbon species has its own characteristic form of the duet. In the siamang, for example, male and female overlap so that both are heard at the same time. In other species, male and female alternate in a sequential duet.

This ritualised calling behaviour leads to the female gibbon's 'great call', of which the pileated gibbon *H. pileatus* has a particularly impressive example. The duet begins at some time early in the morning when the female gives a soft 'hoo-hoo-hoo'. Then, the male and the female hoot in unison in a strange rhythmic call – 'ooh-a-ooh-a-ooh-a'. Eventually, the female utters some short 'hoos' that tell the male to be quiet; this is the signal that she wants to sing the long series of whoops of the spectacular 'great call' – a kind of south-east Asian yodel. At the climax, male and female swing around in the tops of the trees, making a considerable commotion – it is a period of intense excitement.

This strange sequence of vocal behaviour has puzzled naturalists for many years. Warren Brockleman of Mahidol University, Bangkok, has been finding out why gibbons use these complex calls. A primary reason must be to declare and defend territory. The song must be intended for neighbouring groups or why would it be so loud? The long-distance announcement is important for gibbons because physical confrontation might easily end with a damaged arm or hand, and prevent the injured animal from travelling through the branches.

Gibbon songs are clearly a way of avoiding a fight. They are a first line of defence.

An intruding animal can listen to a song and assess the emotional state, physical properties and status of the gibbon making the sound. If the intruder ignores the 'keep out' signal, the resident has a second line of defence. It will confront the intruder at the territorial boundary and 'stare' at it. If that fails, then the resident will chase the intruder. Only when all these lines of defence have failed will gibbons attack one another, and that is a very rare occurrence.

In addition, song analysis by Elliott Haimoff and David Chivers has shown that there are elements in the signal that indicate the age, size, sex and identity of the singer. The degree of synchrony of the duet indicates how well established a pair is in that particular territory. An older, well established pair, for instance, will sing a highly coordinated duet. A newly-mated pair, which has not had time to adjust to each other's idiosyncrasies, will sing with a larger number of errors in the duet.

Warren Brockleman feels that gibbon calls and duets may fulfil other functions as well as territorial proclamation. While a duet is in progress, other groups seldom reply, and the duet is anyway more complex than it need be for a simple 'keep out' signal. It also repeats itself exactly, over and over again. All of these suggest that it is unlikely that the call is only advertising territory. Furthermore, gibbons are supposed to be intelligent primates, so why should they sing a song which is so stereotyped, much like bird song?

One answer may lie in the relationship between the gibbon pair. Like indris and titis, gibbons are monogamous, pairing for life. Indeed, they are so faithful that male gibbons have been seen to chase away other intruding females. David Chivers suggests that duetting in the forest, where visual contact is minimal, might reinforce the pair bond, just as with the Madagascan indri. He likens it to talking to one's spouse at the breakfast table. Many marriages break down for lack of communication. The gibbons' duet is important for the cohesion of the pair, and is also broadcast to neighbours. The strength of a social unit, and its stability, Chivers suggests, is enhanced by isolation. Long-range signals prevent pairs of gibbons continually bumping into each other and disrupting each other's well-ordered lives, and they promote the security of the social group.

Duetting is confined to mated pairs of males and females. Males do not sing with other males, although they sometimes sing at each other across a territorial boundary. Young gibbons sometimes join the parents' duets and so produce a family chorus. Juvenile females will sing in precise synchrony with their mothers, but never alone, until they have left the family and settled with their own mate. It is not known why young females sing in this way. Maybe the maturing female is learning to sing the song in the proper manner, with the correct timing and in the right context. She may, on the other hand, be helping the parents to create a noisier signal for territorial proclamation.

Another aspect of gibbon calls that puzzles Warren Brockleman is why gibbons of different species have such distinctive and easily recognisable calls. Every species of gibbon can be recognised instantly by the form of its duet. This is odd, because the different species occur in quite different areas, separated by major river barriers and straits. The different calls seem to have developed while each species was isolated from the others. There is no selection pressure to discriminate species specific calls, and there is no other animal with the distinctive sound of a gibbon.

Prior to the full-blown song there is a warm-up period when male and female seem to organise themselves for the singing bout. The male siamang, for example, according to Elliott Haimoff, actually invites the female to join him in song. In essence, the male asks the female if she is ready to make her contribution to the 'great call'. She replies either that she is ready or not ready. If she is not ready the male will continue to repeat the request until she indicates that she is ready. The male acknowledges the acceptance and indicates for her to go ahead. He will then join her at an appropriate point. The 'great call' sequence is the result of negotiation between individuals in the pair. Siamangs often groom and occasionally play, so song is but one mechanism which keeps the pair together. Other species of gibbons groom and play less but sing more often. With them, song assumes greater importance in promoting the pair bond.

After the 'great call', there is usually a short pause and then the call is repeated. This goes on for about 15 to 20 minutes. How do they know when to stop? Apparently, there are sound cues for closing the song bout. If one mate wants to stop, but not the other, then the remaining singer will simply continue its part of the duet. If they both want to stop, because of fatigue for example, there are sound signals given by both. It is like two people in conversation on the telephone. To stop the conversation, one of them must give a verbal cue like 'Well, OK', and this would initiate, perhaps, the sequence of 'goodbyes' and each party would hang up. Elliott Haimoff thinks that this kind of human parallel may have more relevance to our understanding of the nature of animal language than one might think. He believes that the fundamental rules and regulations governing the way vocal communication is used may be common to all animals. Take, for example, the way we take turns in speaking during a conversation. The interval between the end of one speaker's turn and the start of the second speaker's turn can be measured in minute fractions of a second and yet, in normal human conversation, only in 5% of cases are there overlaps. There must, therefore, be some kind of pitch cue or other feature in the conversation which allows the other person to know that the first person is going to stop. The same is true in the gibbons. Although the siamang has an overlapping duet, the rest are sequential singers. The interval between one mate's contribution and the start of the second mate's is less than three-tenths of a second. Rarely is there any overlap.

In conversations, humans make errors so there must be a repair mechanism for realising the mistake and then going back over the sentence to put it right.

The gibbons, and several species of birds, have the same mechanism. In the course of singing their complex songs they make errors – a 'frog-in-the-throat', fatigue etc. much like humans, or maybe an animal misses a cue given by its mate. The animals can spot the mistake and go back to correct it, starting once again at the beginning of a given sequence.

In human speech, somtimes, certain sequences of words are always followed by the same response – in ceremonies, for example. When the vicar asks 'Do you take this woman to be your lawful wedded wife?' the reply expected is 'I do'. Any other answer would upset events and a reappraisal of the words might ensue; and maybe the entire ceremony terminated. The same is true of gibbon communication, in that only certain sounds will satisfy particular conditions. The production of sounds other than those specifically designed for the sequence will result either in the sequence being aborted, when the animals will start again at the beginning of the sequence, or in an attempt to repair the mistake and carry on. There are, of course, times when a repair is not desirable. In human terms, when giving a speech, it is often better to blunder on rather than to go back and get involved in a circuitous series of qualifications. For solo gibbons, there is no partner spotting mistakes, so the animal will continue on without trying to correct itself or going back because it wants to carry on with the flow of the song bout.

15

MONKEY BUSINESS

When a dominant vervet monkey *Cercopithecus aethiops* meets a subordinate it grunts. When that same monkey spots a rival troop coming over a distant hill it grunts. If it comes across a tree containing a favourite food it grunts. To the human ear, each of these grunts sounds identical to the others. To the monkeys themselves the grunts are quite distinct in sound and meaning.

Monkey Repertoires

It is the Old World monkeys that have received most attention from the enquiring minds of researchers. The numerous cercopithoid monkeys are long-tailed, tree dwellers living mainly in African forests, south of the Sahara. They inhabit different vertical layers in the forest canopy. The blue monkey or diademed guenon *Cercopithecus mitis* is found chittering its way through the tops of the trees, maybe mobbed by agitated square-tailed drongos and forest weavers. The red-tailed monkey *Cercopithecus ascanius* spends the night in the safety of the foliage and forages on the ground by day. Another group, the thumbless colobine monkeys, *Colobus spp.*, might be startled and begin to bark. The cercopithoids and colobines together with the mangabeys *Cercocebus spp.* are often found living in the same area, a fact which Peter Marler has taken full advantage of. Peter Marler was at Berkeley in 1964 when the Department of Anthropology was developing a strong programme in the study of primate behaviour under the guiding hand of Professor Sherwood Washbourne. Marler went to Uganda to study forest monkeys, in particular, to the Bdongo Forest with its blue monkeys, red-tailed monkeys and colobine monkeys. In those early days, researchers were trying to document the general vocal repertoire and attempting to form some impression of the functions that it served. The rain-forests turned out to be a difficult place in which to work, partly because the animals could not get used to scientists with tape-recorders. Marler did, however, catalogue a number of the obvious calls.

The most conspicuous monkey sound is the 'loud' call. It is a call unique to adult males and is used for long-distance signalling. In order to project the sound, many of the cercopithoid and colobine monkeys have evolved a sac, opening to the larynx, which serves as a resonating chamber for long-distance calls. The 'loud' calls are specific to each species and easily identified.

At the other extreme there is a set of soft, grunting sounds which are used by all animals for maintaining contact between members as the troop moves through dense undergrowth. One call that interested Marler was the alarm call. In comparing the calls used by adult males for spacing out adjacent troops and calls used as danger alarm signals, Marler noticed that, when monkeys talked to members of their own species, the calls were distinct, whereas danger signals were similar from species to species. Different species of small monkeys are vulnerable to the same predators and gain some protection by having signals that are similar enough to allow individuals to talk across species barriers.

In the early days of research, it was thought that a particular type of vocalisation could be linked directly and simply to a particular piece of behaviour. Then researchers began to notice that what seemed to be the same call is given in a variety of situations. A chutter, for instance, might be given when two troops meet, and also when a tree containing a favourite fruit is found. Are the monkeys using the same chutter, perhaps a simple emotional outburst, for both events, or are they, in fact, emitting two quite distinct chutters, the subtlety of which we humans cannot appreciate?

A small New World monkey, related to the marmosets, the cotton-top tamarin *Saguinus oedipus* provided some clues. It lives in the tropical forests of South America, where its numbers are rapidly declining. Indeed, it seems likely that the cotton-top tamarin will survive only in captivity. In order to ensure that breeding populations do thrive in zoos at least, Charles Snowdon of the University of Wisconsin has been studying tamarin behaviour. In looking at the vocal repertoire, Snowdon first noticed that it is complicated. Cotton-tops call for much of their active lives. This is probably related to their habitat. In dense forest they can seldom see each other, and although they have equally complex olfactory signals, vocalisations are an important means of communicating. In the field, Snowdon noted two basic types of calls – chirps and whistles, but on returning to the laboratory and analysing the sounds spectrographically, he discovered more variations. The visual print-outs of cotton-top sounds revealed, for instance, eight different types of chirps, each of which was found to be used under quite different circumstances. There is an alarm chirp given when something disturbs them. An intergroup chirp is heard when contact is made with another group. Another chirp is used when approaching food, and a quite distinct chirp given when the food has been taken and the monkey is walking away.

Snowdon turned his attention to the 'loud call'. Field observations showed that this was given in three situations. When two groups come together, the dominant males emit long calls which seem to function by keeping the two groups apart, thereby avoiding a fight. A second function becomes evident when an animal is lost or becomes isolated. The call is used in order that the rest of the group may find the caller, which is then able to rejoin its lost companions. Thirdly, a group travelling through the forest gives the 'loud call' to facilitate group cohesion. Snowdon found that, instead of there being one call for all three functions, each pattern of behaviour was accompanied by its own subtly different variation of the 'loud call'. By listening to a 'loud call', Snowdon and his colleagues were soon able to recognise an individual and work out what it was doing when it called. Playback experiments with the three different 'loud calls' confirmed the findings.

Switching continents once more, across the Pacific to Japan, another monkey's vocalisations have been under scrutiny. The Japanese macaque *Macaca fuscata* has even more subtle variations in its calls than its South American cousins. Steven Green, of the University of Florida, was one of the first researchers to document the vocal behaviour of this primate. One of the properties of the vocal repertoire, which makes it difficult even to describe, is the fact that the sounds of macaques don't fall into distinct categories but form a continuum, with only very slight differences between them. Are these gradations simply random variations or do they, in fact, have different meanings? Green carried out a careful, quantitative analysis of Japanese macaque calls and discovered that each subtle variation, indeed, does have a different meaning.

One call system, the 'coo calls', was intensively studied. 'Coo calls' are used by macaques when positive contact between individuals is sought. Ten different variations of the 'coo call' were found. A monkey separated from the troop, a mother re-establishing contact with her infant, and a pair in consortship give 'coo calls', all of which are slightly different in structure. Humans can only recognise these differences after spectrographic analysis; so do the monkeys themselves appreciate the structural variations? In collaboration with William Stebbings of the University of Michigan, experiments on sound discrimination in Japanese macaques were carried out. One thing became clear very quickly and that was that Japanese macaques only understand the calls of other Japanese macaques. There is, therefore, some conformity in the variations of calls so that all members of the species attend to the same cues, thereby maintaining common ground in the rules used for communicative behaviour.

Another question tackled was whether the Japanese macaque, or any other monkey, inherits the graded series of vocalisations or whether they are learned. Early work suggested that the production of the sounds is genetically determined but the context in which it is to be used is learned. Tom Strusacker, of the New York Zoological Society, working in Uganda, found support for a genetic base in the production of calls. He discovered troops of blue monkeys and red-tailed monkeys that happened to live closely together. In one part of the forest the two species have interbred producing hybrids. Three hybrid monkeys were located, each living in a red-tailed monkey group.

It was assumed that they were born, bred and raised in a red-tailed society. One of the three, however, an adult male, gave calls that were much closer to blue monkey calls, although he had been brought up with red-tails. This seems to be fairly clear evidence that the genetic component is strong.

Further evidence came from a search for geographical variations. If a call is learned from parents or neighbours, geographically separate groups would develop, over time, their own versions of the basic calls. Strusacker, in examining various populations of monkeys throughout Africa, found that the basic structure of a sound was stable. Comparing the calls of vervet monkeys, for example, in Senegal and Kenya, populations that have separated for several hundreds, if not thousands, of years, the context of calls may differ slightly, but the physical structure is almost identical.

The evidence from field observations seems unequivocal, but when Japanese macaques were brought into the laboratories of Pat Kuhl at the University of Washington, Seattle, doubts began to set in. Several sets of macaque babies were raised under different conditions. A control set was reared with a family group in as normal a situation as can be devised in captivity. Other babies were taken from their mothers at birth and raised with other infants of their own and other species. A third group were raised away from adults but were played adult calls over loudspeakers. At the time of writing, results had not been obtained from the third group. Observation of the other two groups showed that there are dramatic differences between the sound production infants make without hearing adults and those exposed to adult example. The calls given by infants that have not heard conspecifics seem to be emitted less frequently and are uneven. Kuhl has demonstrated that the production of normal vocalisations in Japanese macaques is dependent on early auditory experience. This is a learned component; a new result from the standpoint of perception development in non-human primates.

Vervet monkeys

Vervet monkeys *Cercopithecus aethiops* live in troops, mainly in the open savannah areas of Africa, south of the Sahara. They are medium-sized monkeys with a greenish hue, black face, white throat, and white whiskers. The sounds that they make have great scientific interest, for each sound is thought to correspond to a specific object or event in the external world – a property usually reserved for human speech.

A detailed analysis of vervet monkey sounds is being carried out by husband and wife team Dorothy Cheney and Robert Seyfarth, of the University of California at Berkeley. Their interest was triggered by recent developments in ape sign language and other forms of communication in higher primates. Cheney and Seyfarth believed that vervets might be a useful living tool with which to investigate whether species other than man have evolved the ability to make systematic use of signals to indicate objects around them. The clue was that vervet monkeys have distinct alarm calls for different predators.

It was while working with the vervet monkeys in the Amboseli National Park, in the mid 1960s, that Tom Strusacker made the initial observations. He noticed that if the monkeys were foraging on the ground in an area of open bush, and one saw a leopard approaching, it would give a loud bark and all the rest of the troop would flee to the trees. If a monkey spotted a martial eagle or a crowned hawk eagle soaring overhead the 'troupe' alarm call would be quite different, more of a chuckle than a bark, and the others would look up into the air or dive for cover into a bush. The monkeys' responses to these classes of alarm calls relate closely to the hunting strategies of the predators. Leopards hunt during the daytime, hiding in bushes, waiting to ambush monkeys passing by. A leopard can take a vervet on the ground but is not quick enough in a tree. An eagle can take vervets, both on the ground and in trees, so the safest place to avoid eagles is in dense bush.

When the troop came upon a snake, particularly a python, a high-pitched chittering alarm would cause the monkeys to get up onto their hind legs and look carefully around them; sometimes they would mob the snake. There are also separate and distinct alarm calls for baboons, which sometimes attack young vervets, and of course alarm calls for man, sometimes the most feared of all predators.

Dorothy Cheney and Robert Seyfarth decided to investigate the specificity of meaning of the alarm calls. They carried out a series of experiments, in the field, in 1977 and 1978. They located a troop of vervets foraging in a particular area and planted a loudspeaker in a bush nearby. They were then able to play back various monkey alarms recorded earlier in natural encounters with leopards, eagles and snakes. When they played the alarm calls the monkeys reacted predictably. Cheney and Seyfarth filmed the events on a Super-8 sound movie camera and, back in the laboratory, were able to replay over and over again, with synchronised sound, the monkeys' behaviour before, during and after the alarm call was broadcast from the loudspeaker. With a playback of the leopard alarm, all monkeys raced for the trees. With the eagle alarm they looked up and moved to dense bushes. The snake alarm evoked the snake-hunting behaviour. What Cheney and Seyfarth had shown was that the alarm call, in itself, even in the absence of the actual predator, means something sufficiently specific to the monkeys to function as if it were a word, thus dispelling the traditional notion that an animal signal can be only a general manifestation of fear. It may be that an eagle alarm, for example, may refer to a recognised bird of prey, a collective noun for flying predators, or a particular escape instruction, but the important thing, as far as the Seyfarths are concerned, is that the sound represents a particular object or event in the external world.

Another question Cheney and Seyfarth tackled is whether the alarm calls are innate or learned. How do the alarms develop and how is their use modified during adulthood? Here there are two facets to consider – the production or phonetics of the calls and the use to which they are put. As yet the production question has not been considered. There is no data on whether a youngster's alarm call differs in structure from that of an adult. There are, however, interesting observations about the circumstances in which these calls are given by juveniles. When adults give leopard alarms the call is almost exclusively for leopards, although lions, hyaenas, cheetahs and jackals are greeted with the same call. Martial eagles and crowned hawk eagles were the main triggers for the eagle alarm, but black-chested snake eagles and tawny eagles, both potential predators, elicited the eagle call. The snake alarm, primarily for pythons, was given also for cobras, black mambas, green mambas, and puff adders.

Infants give alarm calls to all sorts of visitors that don't prey on the monkeys. They give leopard alarms to warthogs, hawk alarms to pigeons, Egyptian geese, African goshawks, grey herons, ground hornbills, lilac breasted rollers, Marabou storks, secretary birds and spoonbills, and snake alarms to harmless snakes, tortoises and even mice. One individual gave a hawk alarm to a falling leaf. The alarm calls, however, even from the earliest days are not random. Hawk alarms are given only to flying objects. Leopard alarms are given only to animals that walk along the ground, such as wildebeest, zebra and hogs. Snake alarms may be given for all sorts of snake-like objects such as vines, as well as to non-poisonous snakes, poisonous snakes and pythons. Infant vervets clearly start with some kind of predisposition to give alarm calls to broad classes of predators and subsequently narrow the field down. Adults appear to teach the juveniles which objects warrant which alarms. If an infant gives an eagle alarm to a pigeon, the adults nearby will look up into the sky, see the harmless bird, and do nothing. But, if the infant gives an eagle alarm to a martial eagle, the adults will look into the air, see the eagle, and give the eagle alarm themselves. The adults' response reinforces the infant's behaviour; the infant is told either 'You've got it right' or 'ignore pigeons they're safe'.

Cheney and Seyfarth tried playback experiments. They waited until a youngster became separated from its mother and then played an alarm call. A very young infant would be startled and confused, looking this way and that, searching for its mother to know what to do. Finally it would run to the safety of the parent. The adults, though, have been giving the infant many cues about the way in which it should behave. A few days later Cheney and Seyfarth would play an eagle alarm to the same infant and film its response. A frame-by-frame analysis revealed that the infant would look to the mother, see her looking up, and then look up into the sky itself. A few weeks later the juvenile would respond to an eagle alarm just by looking up.

So, as they grow up, infants appear to sharpen the association between specific classes of predators, the specific alarm calls, and the responses to those calls. Furthermore, the monkeys seem to have some ability to change both their alarm calls and their responses depending on the hunting behaviour

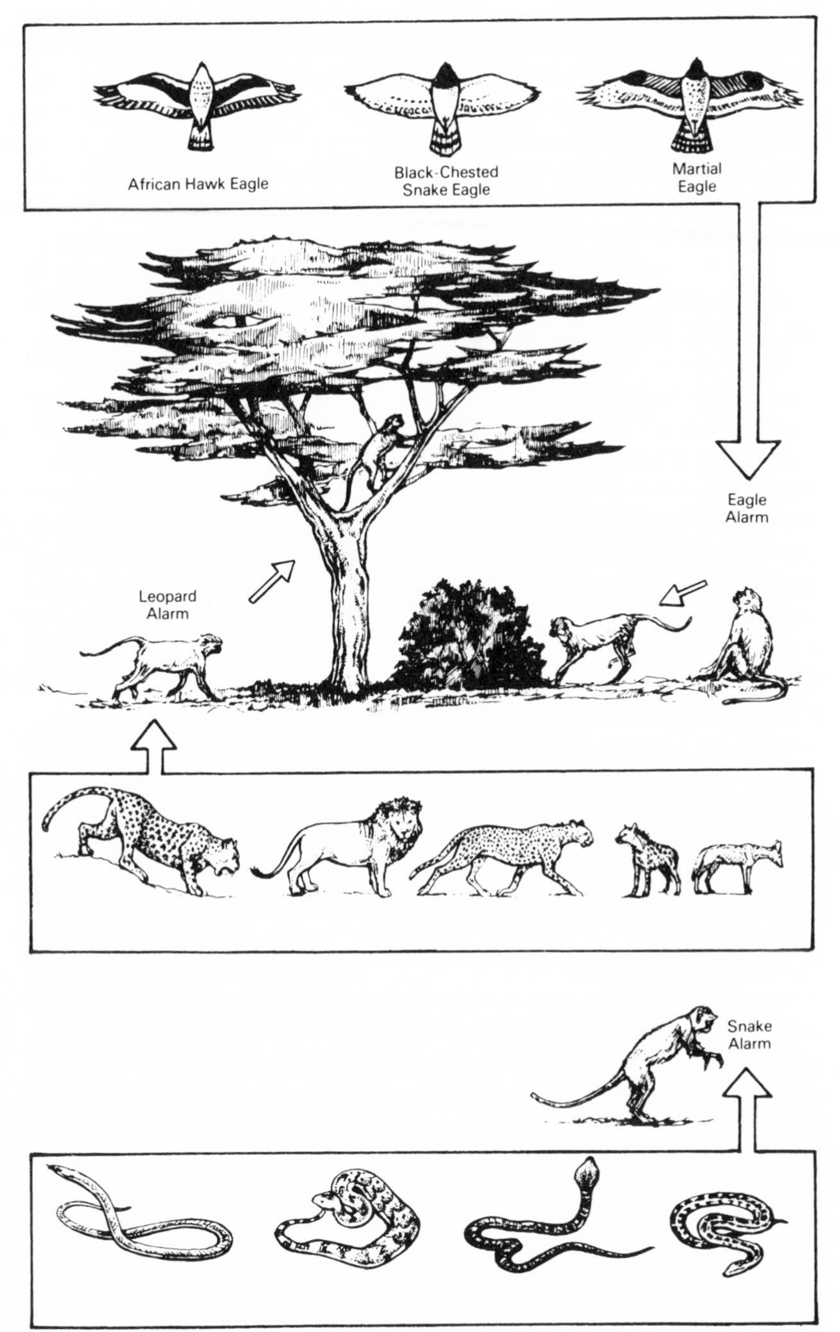

Fig. 47 Response of vervet monkeys to specific predators.

of their predators. In the Cameroon forests, for example, where vervets are hunted by humans with dogs, the monkeys' alarm calls are soft and pitched within a frequency band that matches the ambient background noise of the forest, making them difficult to detect. The monkeys respond by fleeing silently into dense bush where humans and dogs cannot follow. In nearby savannah habitats, however, where vervets are not hunted by man, the monkeys climb quickly to the tops of trees uttering loud alarm calls.

Loud alarms, such as those given by vervets, are likely to attract the attention, not only of the animal's companions, but also of the predator itself. Could this then be an example of altruism? Robert Seyfarth thinks not. Vervet monkeys are good at keeping a look-out and are likely to spot approaching predators long before the danger is acute. Eagle alarms, for instance, are given where eagles are many hundreds of yards up in the sky. Vervets, unlike ground squirrels, are not primarily concerned with the protection of offspring. As often as not, the predators of adults are quite different from those of juveniles. Adults give alarm calls to animals that threaten themselves. Leopards and eagles prey mainly on adults. Baboons take young vervets. If a baboon approaches with intent to kill, an adult's alarm call does not put a large adult vervet at risk. The adult vervet is too big to need to be worried. It is the juveniles that need the information. Consequently large adults rarely give alarm calls for baboons, yet the juveniles do so. Youngsters give fewer alarm calls to leopards and eagles than they do to baboons. Adults give alarms to protect adults and juveniles give alarms to protect juveniles, despite the fact that an adult crying out to protect a juvenile would expose itself to little danger. There is, however, some evidence to show that the adult males that give the most alarm calls are those with most offspring in a group. The same is not true of adult females, although the offspring of high-ranking females are more prone to attacks, as dominant animals tend to lead the troop across the savannah. Cheney and Seyfarth believe 'that both kin and individual selection acting on an individual's inclusive fitness, have played a role in the evolution of vervet monkeys' alarm calls'.

By watching vervet mothers' responses to their offspring, however, Cheney and Seyfarth were able to make some remarkable observations. Vervets have a well defined and complex social system. Most groups of Old World monkeys have in their care a number of genetically related adult females and their offspring. The females form a linear dominance hierarchy, which is reflected in the social organisation of the youngsters. Each infant acquires rank according to its mother's position in the troop. When two infants fight, one might scream. The mother comes running to investigate. Her ability to defeat or chase away her infant's rival is a direct function of her social status in the troop. An infant gets the rank its mother earns for it.

Cheney and Seyfarth noticed that mothers could recognise the screams of their offspring. They waited until three resting mothers were seated together. The scream of one of the mother's infants was played and the response filmed.

It was not surprising that the mother responded more quickly and strongly than the other two females by looking towards her infant. What did surprise Cheney and Seyfarth was that the other females responded by turning instantly towards the mother. The mother vervet's reaction had shown that she could discriminate her own offspring from other vervets. The control females had demonstrated a lot more. They showed that vervet monkeys not only identify individuals but are capable of recognising the individuals as members of different families.

Has Cheney and Seyfarth's study of alarm calls, though, really shown that vervet monkeys are capable of word-like communication? Many birds and mammals have alarm calls that differentiate particular groups of predators, yet no-one is arguing that robins or great tits, for example, have a linguistic system. Do vervets, in their other types of vocal communication, also have sounds that signify something specific external to the signaller? To examine this Cheney and Seyfarth have turned their attention to the most common vocalisation of the vervets, the grunt.

Vervet monkeys grunt to each other in a variety of circumstances. On moving out of their sleeping trees each morning at dawn vervets begin to forage. One animal may approach another that is more dominant and give a grunt (like someone clearing their throat with the mouth open). Fifteen or 20 minutes later the same monkey may approach a subordinate and again a grunt is heard. At the edge of a forest clearing, the same monkey sits with its companions waiting to cross in the open. The troop is nervous. The monkey grunts again. A little later the group is foraging; the monkey looks up and grunts. Following its line of gaze reveals another troop on the horizon.

All the grunts given by the vervets sound exactly the same to human ears. Traditionally this would be considered as one vocalisation, an uncontrollable manifestation of nervousness – one monkey grunts, another looks up, checks around and sees the other troop: the single grunt could mean a number of different things, and is not in itself like a word but more of a general signal.

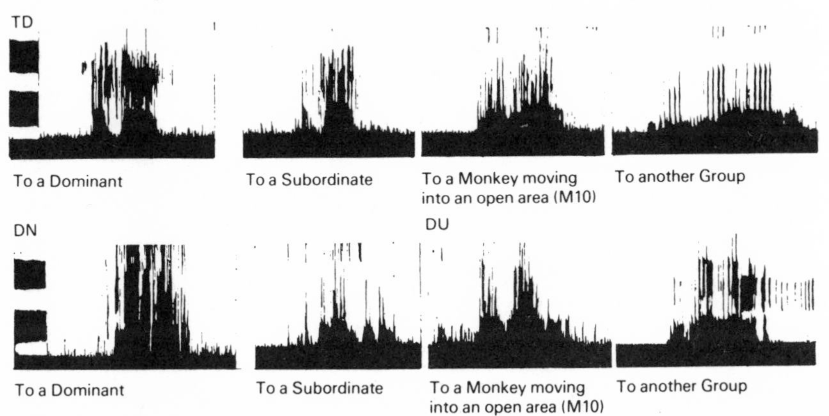

Fig. 48 Vervet monkey grunts.

Cheney and Seyfarth considered this traditional view unsatisfactory. If monkeys have the ability to use different alarm calls for different predators, then why not different grunts? Unfortunately, the human ear cannot discriminate grunt variations. Robert Seyfarth likens this predicament to that facing social anthropologists studying, say, African tribes using click language. There may be ten different clicks but western observers can only hear one. Cheney and Seyfarth attempted to reveal variations in grunting behaviour in the wild by getting the monkeys themselves to tell them whether they were using just one grunt or several different grunts. Cheney and Seyfarth again used playback experiments. Grunts were recorded from monkeys in a variety of circumstances. A vervet foraging alone was located and a loudspeaker placed in a nearby bush played a grunt. Three or four days later they played a different grunt and so on. Cheney and Seyfarth were working through the monkeys' dictionary, presenting word one and watching to see if it led to a different response to that covered by words two, three and so on. They were able to show, for instance, that a grunt from a subject approaching a dominant individual consistently evoked a different response – a turn of the head, a different posture, a different gaze – from a grunt originally recorded when approaching a subordinate. A vocalisation perceived by human ears as one grunt is received by the monkey as many subtly different calls, each one associated with a specific and entirely different set of social circumstances. Dorothy Cheney and Robert Seyfarth, at present, continue to try to unravel vervet monkey sounds, having established the first few entries in the English-vervet dictionary.

16

ANIMAL LANGUAGE?

It might seem odd to question the title of the book in the last chapter, and return to the question, do animals have language? Could all those squeaks, clicks and burps be candidates for language? Humans have whistle and click languages, so it need not be that the formation of complex words is important. A tone of voice does not count as language; rather it is an emotional expression, although it may have been learned from parents or in nursery school. An outburst, like a scream accompanying a fall down the stairs, is not thought to be a component of language. Indeed, there appears to be a complete parallel system of non-verbal communication, even in human voices. Here there is a clue. One feature of non-verbal communication is that it is hard to lie. Actors can fake it, but generally it is a reflection of your feelings, a good barometer of emotions. With language you can lie. It is all right to say, 'I am a big red bus', even though you are not, because it is still a perfectly good sentence. This separation of meaning seems to be a good criterion of language. In addition, with language you can tell of something not obviously visible such as a wildebeest or a field of berries on the other side of the hill, or even tell of something taking place in the future or recall something from the past. It is an advantage, in evolutionary terms, to be able to communicate something that is out of sight but not out of mind. When an animal can make the jump to talking about something that isn't there we would say it has language. Very few animals show signs of being able to do this. Many of their sounds are closely tied with their emotional state. Indeed, even in humans most communication is non-verbal communication, or what J. A. R. A. M. van Hooff calls 'grooming talk', which has very little to do with words – 'please pay attention to me, I'm noticing you'.

Some animals, though, do have the equivalent of human words. Robins distinguish between aerial and ground predators and give specific alarm calls depending on which predator is spotted. Ground squirrels respond differently to a hawk alarm call from a badger alarm. Any burrow will do to escape the hawk, but a burrow with a back-door is the only safe refuge when a badger is about. Perhaps the most interesting case is that of the vervet monkeys with their distinct alarm calls for leopards, eagles, pythons, baboons and man. Some people argue that these simply represent a gradation of emotionally induced alarms responding to increasingly more dangerous threats, although why a hawk, say, should be considered more of a threat than a leopard is not clear. It is likely that these alarms are, indeed, the equivalent of human words, representing objects in the outside world by separate and discrete sounds. But, could vervets take this further? Could they give a snake chutter having come away from the snake and inform other vervets of a danger that they might encounter 'round the next bend'? At present we don't know, but there is one primate, other than man, that appears to be able to do just that – it is the chimpanzee, but it doesn't do it with just verbal signals.

E. W. Menzel carried out a series of experiments, in a large field, with half-a-dozen captive chimpanzees. He would take out one or two at a time, show them a pile of food hidden under a bush somewhere in the field, and then return them to the other young chimps. The group preferred to go out into the field together. They wouldn't dare go out alone to find the food, so the two that knew there was food out there had to tell the others that there was something worth going out for, and encourage them to follow. If the scouts were dominant animals they would lead out into the field and the others would follow. If the first pair were subordinate then they would pull the fur of the others and look beseechingly towards the food. If they were ignored they would throw tantrums.

Menzel then complicated the experiment. He would take two chimps out to see two different piles of food. Sure enough, they decided unanimously to go for the larger pile, no matter which chimp was shown which pile.

In further tests Menzel used objects that terrify chimps, like snakes. He would hide a rubber snake under some leaves in the field, show one chimp the spot, and then return him to the others. This time they behaved quite differently. They would creep up with hair bristling, fear grins on their faces, slap with their hands on the leaves until the toy had been uncovered, and then beat it to death.

Menzel went even further. He showed the chimp a snake, hid the snake in the leaves, but then took it away. The chimps' reaction to this was even more dramatic. They would approach, as before, to the spot where the snake had been hidden, discover it was not there, and then systematically look up and down the entire field until they were satisfied that the danger had gone.

In some way Menzel's chimpanzees must have told each other that there was food worth getting or a dangerous object which might threaten the safety of the group. We know that chimpanzees have the capacity for symbolic memory, for instance, as shown in sign language experiments, so could it be that chimpanzees have reached the stage where they have the capacity to think ahead and make a mental image of something that is not readily visible? Some researchers believe this is the case.

The sign language experiments are of particular interest to researchers working on sound communication, in that they give some clues about the language abilities of primates, although these laboratory based experiments have their limitations. Over the past five to ten years, several researchers have demonstrated that captive apes can be taught a rudimentary linguistic form of signalling. Despite the controversy surrounding this work, it has shown that apes are able to learn a large number of single words. They can learn the relationship between, say, a plastic token and a sign. They can also relate to an object seen the previous day but not in front of them at present.

If, then, these apes in the laboratory can be taught to perform this language-like behaviour what are they saying to each other in the wild? Somehow they must have evolved this ability. Natural selection must have favoured some sort of cognitive ability that has now become apparent to us, albeit in an unnatural setting. There are, however, problems. Herb Terrace, reviewing the sign language work, has drawn attention to the fact that when an ape fails to perform a task, when it can't, for example, combine several words into a sentence or use a grammatical structure, we do not know whether that represents a failure of ability, i.e. the animal is not smart enough, or whether it represents a failure of motivation. An ape performing in the human world is asked to learn a human form of signalling. Using that signalling, it is interrogated in human-style interview interactions. This is like taking a Frenchman and asking him questions in Finnish. The subject is not going to give his best performance under these conditions. The thrust, then, of Dorothy Cheney and Robert Seyfarth's work with vervet monkeys in East Africa is not to teach apes our language, but to tackle the apes and monkeys on their own terms. In this way the researchers go into the monkeys' and apes' own habitat, learn *their* form of communication, and ask them to tell us, in their own words, what they are signalling about. In this way the Seyfarths hope to complement the ape sign language tests, and possibly shed some light on the problems that these projects have so far failed to resolve; for example, there is no evidence that a chimpanzee can take a group of single words, which it knows how to use, and combine them into a sentence or a compound word – a word that is more than the sum of its parts. To this end the Seyfarths are focusing on the vervet monkeys and their grunts. So far, they have discovered that although the social grunts sound the same to our ears, the monkeys are, in fact, giving subtly different grunts in different circumstances. Could it be that, when a monkey gives a string of grunts, it pays attention to grunt order? Or, might a monkey combine two grunts to make an intermediate that has a fresh meaning, more than the sum of its parts? The Seyfarths are now tackling this frontier in their research, in the hope that by investigating the monkeys on their own terms, where there is *need* to signal, they will get to know the full extent of vervet monkey abilities.

Why, though, has language evolved? Why is it that humans use this extremely complex form of signalling when other animals, as yet, do not appear to be capable of it? One way that natural selection could be considered to work is to say that 'necessity is the mother of invention'. We have this complicated form of communication because we need it. At some stage in the distant past an individual, by combining words into sentences, gained an advantage over the rest in his group so that all the others had to learn it as well. So, what advantage would a vervet monkey gain if it could take a bunch of grunts and string them into a sentence? Do vervets, in fact, do this, and if so, what is the likely social area in which they do it? And, perhaps, a more important question, do they need to do it? Man is clearly not alone in being able to use symbols to represent objects or events, but is there a need for an animal to go any further? Possibly there is not. In many of the situations in which animals need to communicate they do not have the time to construct complicated signals, strung out in long series. Information transfer has to be rapid in the many life and death situations that arise in everyday life.

On the other hand, animals need not join their words end to end. It could be that animals construct simultaneous, rather than sequential, sentences, with a number of components, each with its inherent meaning, combined as an overlay and broadcast together. This could be one reason why we have not recognised sentence construction in animals. Failing such a discovery, could it be that human language is the Rubicon that divides humans from the other animals?

But might animals give us some insights into the origins of our own utterances; after all, many of the primates being studied live in a similar environment to that thought to have been inhabited by early man when he swung down out of the trees? Vervets live on the forest edge; macaques, baboons and some groups of chimpanzees live on the open plains. The clue is that these animals of the savannah are in tight social groups. Non-human primates living on open grasslands are always in groups. In forests there are to be found solitary monkeys. They are still social but there is no taboo on living alone. Rarely are there to be found solitary monkeys on the plains, probably due to predation pressures. In this way early man would have found safety in groups, and in groups individuals must communicate. Possibly the most important thing to a group member is another group member. The most successful members of a group are going to be those that are most able to manipulate their fellows. As Robert Seyfarth once put it, 'group members become scheming Machiavellians in a cauldron of social competition', and the ability to use a vocal signal to manipulate others is going to have a strong premium placed on it; an ideal situation in which a sophisticated language could have evolved.

So, at the end of this book we are left with a tantalising glimpse of how animals, other than man, have developed systems of vocal communication. But it is clearly only the tip of an iceberg. Almost every week scientific journals report new and ever more astonishing abilities of animals, abilities which we never suspected, and many of which we ourselves lack. The study of sound communication in animals has an exciting future. We are only now beginning to reap the benefits from the availability of new techniques for analysing animal sounds. Until sound spectrographs, for instance, were available we could do no better than verbalise sounds and write down imperfect renditions. It has taken 15 to 20 years to develop a sufficient corpus of knowledge about the structure of animals' sounds to prepare researchers for the next phase where they will begin to expect that some of the subtleties and variations that are detectable are not biologically irrelevant but have meaning to the species that use them. Communication researchers are in an explosive phase of their science now where, with a little imagination, it can be guaranteed that, with a careful choice of topic, someone will discover something new and unexpected. Professor Peter Marler feels that researchers are preparing for a quantum leap in their expectations with regard to the complications and subtleties of animal language.

BIBLIOGRAPHY

CATCHPOLE, C. K. *Vocal communication in birds* Edward Arnold, 1980.

DARWIN, C. *The expression of the emotions in man and animals* J. Friedman, n.e. 1979, University Chicago Press, 1965.

HALLIDAY, T. *Sexual strategy* Oxford University Press, 1980.

JELLIS, R. *Bird sounds and their meaning* Cornell University Press, 1984. Sounds illustrated in this book are available on cassette tape from: Crow's Nest Bookshop, Laboratory of Ornithology, 159 Sapsucker Woods Road, Ithaca, New York 14850.

KREBS, J. R. and DAVIES, N. B. eds. *Behavioural ecology; an evolutionary approach* Blackwell Scientific Publications, 1978.

LEWIS, D. B. and GOWER, D. M. *Biology of communication* Blackie, 1979.

McFARLAND, D. ed. *The Oxford companion to animal behaviour* Oxford University Press, 1981.

SALES, G. and PYE, D. *Ultrasonic communication by animals* Chapman and Hall, 1974.

SEBEOK, T. A. ed. *How animals communicate* Indiana University Press, 1977.

Acknowledgements

I would like to thank Anthony Arak, Jeffery Boswall, John F. Burton, Clive Catchpole, David Chivers, Brian Lewis, David Pye, Peter Slater and Peter Tyack for reading sections of the book, and for their invaluable comments, criticisms, suggestions and amendments.

In addition I am grateful to David Aidley, Henry Bennet Clark, Brian Bertram, Martin Birch, Warren Brockleman, Sharron Brownley, Bob Capranica, Chris Clarke, Tim Clutton-Brock, John Deag, Arthur Ewing, Bruce Falls, Brock Fenton, John Ford, James Fullard, Wilma George, Carl Gerhardt, Stephen Green, Don Griffin, Mike Gunton, Eliot Haimoff, Joan Hall-Craggs, Tim Halliday, Tony Hawkins, Michael Jarman, Alison Jolly, Pat Kuhl, John Krebs, Donald Kroodsma, Hans Kruuk, David MacDonald, John Mercer, Eugene Morton, Robert Moss, James Moulton, Ken Norris, Fernando Nottebohm, Roger and Katy Payne, Robert Payne, Michael Petersen, David Ragge, Stanley Rand, Michael Ryan, Gillian Sales, Robert and Dorothy Seyfarth, Charles Snowdon, Tom Strusacker, W. H. Thorpe, George Uetz, Elizabeth Walser, Kentwood Wells, Haven Wiley and Margaret Vince, for their time and help and their fascinating contributions to the series of radio programmes. Without access to their research the series and book would not be possible.

I am particularly indebted to Peter Marler for his patience in the face of my barrage of eager questions; to Tom Eisner for his enthusiasm; to Mary Welland for the initial programme research; and to Monica Simms for her encouragement of the entire project. Kate Tiffin must be congratulated on being able to decipher my scrawl, and Robert MacDonald on being able to recover meaningful sections of the English language.

Picture credits

1 BBC; 2 New York Times (Walter Sullivan); 3 Frank Lane (S. McCutcheon); 4 Ardea; 5 Ardea (R. J. C. Blewitt); 6, 7, 8, 9, 10 Photo Researchers Inc (Merlin Tuttle); 11 F. A. Webster; 12, 13 Brock Fenton; 14, 15 George Uetz; 16, 17 Susan Silver; 18 Barnaby's Picture Library (P. Huggins); 19 Bruce Coleman (Hans Reinhard); 20 Bruce Coleman (Jane Burton); 21 Ardea (Clem Haagner); 22 Jeremy Cherfas; 23 Barnaby's Picture Library (Gerald Cubitt).

Drawings by Maggie Raynor; diagrams by Brian Rockett.

INDEX

Bold numbers refer to plates.

INDEX OF SCIENTISTS AND INSTITUTIONS

Library of Congress Cataloging in Publication Data

Bright, Michael.
Animal language.

Based on the BBC Radio 4, series Animal language.
Bibliography: p.
Includes index.
1. Animal communication. 2. Animal sounds. I. Animal language (Radio programs) II. Title.
QL776.B75 1985 591.59 85-6714
ISBN (cloth) 0-8014-1837-2
ISBN (paper) 0-8014-9340-4